CORONEL **SERIES** 02

Making history!

www.editorialinfinito.com.ve

250 SOLVED EXERCISES FROM DERIVATIVES WITH APPLICATIONS

(c) Editorial Infinito, 2018
(c) Pedro Pablo CORONEL PÉREZ / Pablo Josué CORONEL LÓPEZ

LEGAL DEPOSIT: lf07620145102757
ISBN: 978-179-09-8248-6
INTERNAL LAYOUT: Editorial Infinito / Estudiográfico
COVER DESIGN: Estudiográfico (2018)
LITERARY EDITOR: Magister / Lcdo. Pedro Alberto Coronel López
DIGITAL PRINTING: Editorial Infinito, San Cristóbal

Any observation, suggestions and correspondence are kindly requested to be sent to the following
emails: **pedro_coronel1955@hormail.com / pablocoronel19@hotmail.com**

Produced by Editorial Infinito (República Bolivariana de Venezuela)

CORONEL SERIES

250

SOLVED EXERCISES FROM
DERIVATIVES
WITH APPLICATIONS

[INCLUDES THEORETICAL BASIS]

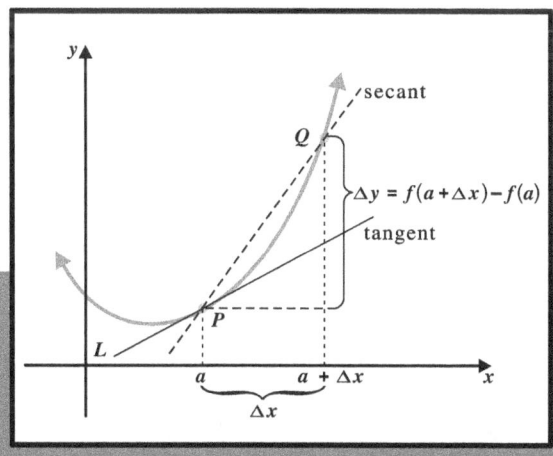

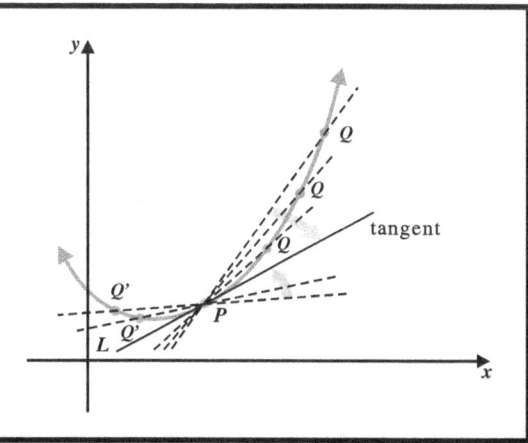

PEDRO P. **CORONEL PÉREZ** | PABLO J. **CORONEL LÓPEZ**

Another titles of the CORONEL SERIES
Published by Editorial Infinito

250 LIMITS RESOLVED OF THE FORMS

$$\frac{0}{0}, \frac{\infty}{\infty}, \infty - \infty, 1^{\infty}$$

To my first grandchild, **Massimo Lorenzo**

"Mathematics is the alphabet with which God has written the universe."

GALILEO GALILEI (1564-1642) / Italian philosopher and mathematician

CONTENTS

PEDRO PABLO CORONEL PÉREZ

Retired professor of the university institute of agroindustrial technology Region los Andes Bolivarian Republic of Venezuela. He was assigned to the department of electronics in electrical circuits and physics laboratory. I collaborate in the national training project in the area of mathematics..

PABLO JOSUÉ CORONEL LÓPEZ

Electrical engineer graduated from the university institute of agroindustrial technology Region los Andes, Bolivarian Republic of Venezuela.

COMMENTS FROM THE AUTHORS

Welcome to the first edition of 250 solved derivative exercises with calculation applications 1! The Infinite Publisher is proud to present the aforementioned book to the student and teacher community. The purpose of this book is to present to those who start university studies, a series of exercises on DERIVATIVES, very representative and resolved in detail. Obviously, it will be very useful for career students linked to engineering, science, technology or any specialty where mathematical calculation is an essential requirement within the study curriculum.

The number of exercises included allows the book can also be used as text by both the student and the teacher in the development of this important subject of calculation.

The authors have been careful in the explanation of the procedures used in the resolution of each of the problems. The exercises have been selected in order to expand the knowledge acquired in class, as well as for the student to acquire practice in solving problems and thus prevent the difficulties with which the learner normally stumbles.

We hope you enjoy the first edition of 250 Resolved Derivatives Exercises with Calculus Applications 1. As always, comments and suggestions to continue improving the work are welcome.

Pedro Pablo Coronel Pérez
AUTHOR

Pablo Josué Coronel López
AUTHOR

Pedro Alberto Coronel López
LITERARY EDITOR

THEORETICAL BASIS FOR DERIVATION

STRAIGHT TANGENT TO A GRAPH

Suppose that y = f (x) is a continuous function whose graph is shown in Figure 1 (a). If the graph of f has a tangent line L at a point P, as illustrated in Fig. 1 (b), the problem is to determine its equation. To do this, it is required: (a) the coordinates of P and (b) the slope m tan of L. The coordinates of P do not present difficulty, since a point of the graph is obtained by specifying a value of x, for example, x = a, in the domain of f. The coordinates of the point of tangency are (a, f (a)).

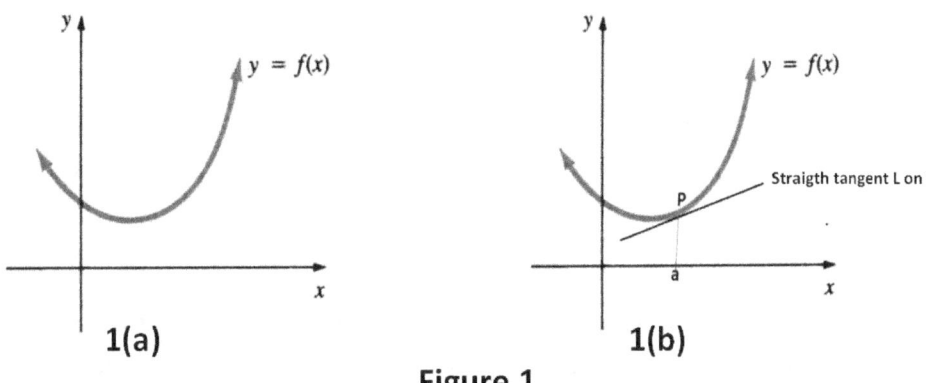

1(a)   **1(b)**

Figure 1

One way to approximate the slope m_{tan} is to determine the descent of secant lines that pass through point P and any other point Q of the graph. If P has, coordinates (a, f (a)) and Q is made by coordinates (a + Δx, f (a + Δx)), then, as shown in Figure 2 (a), the slope of the secant line what happens through P and Q is:

$$m_{sec} = \frac{change\ in\ the\ coordinate\ and}{change\ in\ the\ x\ coordinate}$$

$$= \frac{f(a + \Delta x) - f(a)}{a + \Delta x - a}$$

$$\Delta y = f(a + \Delta x) - f(a)$$

If

$$\Delta x = (a + \Delta x - a)$$

So $\quad m_{sec} = \dfrac{\Delta y}{\Delta x}$

When the value of Δx is small, either positive or negative, points Q and Q' are obtained from the graph of f on each side of point P, but close to it. It is expected that, in turn, the slopes m_{pq} and $m_{pq}{}'$ are very close to the slope of the tangent line of L. See figure (2b).

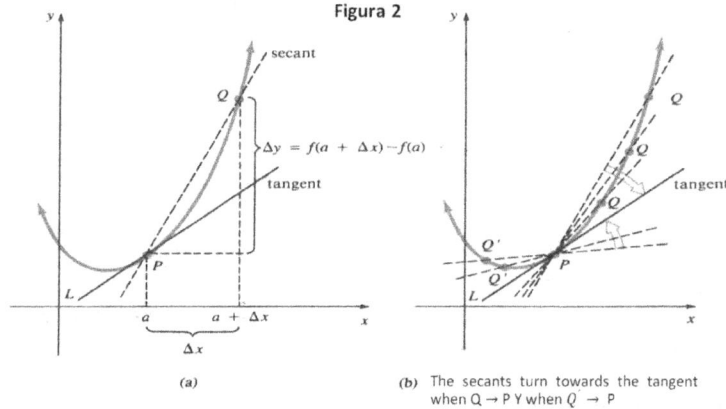

(a)

(b) The secants turn towards the tangent when Q → P Y when Q' → P

Based on Figure 2 (b), it could be said that if the graph of a function y = f (x) has a tangent line L at a point P, then L must be the line that is the limit of the blotters that pass by P Y Q when Q → P, and of the secants passing through P and Q' when Q' → P. In addition, the slope m_{tan} of L must be the limit value of the m_{sec} values when Δx → 0. This is summarized as follows:

DEFINITION 1

Let y = f (x) be a continuous function. The line tangent to the graph in the point (a, f (a)), is the one that passes through the point and its slope is:

$$m_{tan} = \lim_{\Delta x \to 0} \frac{f(a+\Delta x) - f(a)}{\Delta x} = \lim_{\Delta x \to 0} \frac{\Delta y}{\Delta x}$$

Provided that the limit exists.

The slope of the tangent line in (a, f (a)) is also known as slope of the curve at the point. Definition 1 implies that a tangent in (a, f (a)) is unique, since a point and a slope determine a single line.

THE DERIVATIVE

From the above, it was determined that if the graph of a function y = f (x) has a tangent at the point (a, f (a)), then the slope of this is

$$m_{tan} = \lim_{\Delta x \to 0} \frac{f(a+\Delta x)- f(a)}{\Delta x}$$

For a given function it is usually possible to obtain a formula, or general rule, that provides the value of the slope of the tangent line. This is done by evaluating:

$$\lim_{\Delta x \to 0} \frac{f(x+\Delta x)- f(x)}{\Delta x}$$

For any x (for which the limit exists). Then a value of x is substituted after finding the limit. The previous limit is known as the derivative of f and is denoted by f '.

DEFINITION 2

$$f'(x) = \lim_{\Delta x \to 0} \frac{f(x+\Delta x)- f(x)}{\Delta x} = \lim_{\Delta x \to 0} \frac{\Delta y}{\Delta x}$$

Provided that this limit exists. The derivative f'(x) is also called the instantaneous rate of change, of the function y = f (x) with respect to the variable x.

RHYTHMS OR RELATED CHANGE SPEEDS

According to what was studied in the previous section, the derivative allows to perform calculations to find slopes. But it also serves to determine the rate of change of one variable with respect to another. The following examples constitute rates of change population growth, production rates, flow of a liquid, velocity and acceleration.

To describe the movement of an object that goes in a straight line rhythms of change are used.

The movement of the line is usually represented in horizontal or vertical position, with a marked origin in it. On such lines, movement to the right (or upward) is considered positive direction and movement to the left (or down) of negative direction.

The average speed (v_m) or average speed of a moving object is the ratio of change of position with respect to time, defined by

$$v_m = \frac{change\ in\ distance}{change\ in\ time} = \frac{\Delta s}{\Delta t}$$

If s = s (t) is the position function of a rectilinear moving object, its velocity at time t is

$$v(t) = \lim_{\Delta t \to 0} \frac{s(t+\Delta t)-s(t)}{\Delta t} = s'(t) \quad \rightarrow \quad \text{Speed function}$$

In other words, the velocity function is the derivative of the position function. The speed can be positive, zero or negative. The speed of an object is defined as the absolute value of its speed, and it is never negative.

The position of an object in free fall (neglecting the air resistance) under the influence of gravity is obtained by the equation:

$$s(t) = \frac{1}{2}gt^2 + v_o t + s_o \quad Position\ function$$

Where s_o is the initial height of the object, v_o the initial velocity and g the acceleration of gravity, which on the earth's surface is - 9.8 m / s.

DERIVABILITY AND CONTINUITY

The following alternative form is the limit of the derivative is useful when investigating the relationship that exists between derivability and continuity. The derivative of f at x = c (or at x = a), is

$$f'(c) = \lim_{x \to c} \frac{f(x)-f(c)}{x-c}$$

provided that such limit exists.

Note that the existence of the limit in this alternative form requires that the unilaterals limits

$$\lim_{x \to c^-} \frac{f(x)-f(c)}{x-c} \qquad and \qquad \lim_{x \to c^+} \frac{f(x)-f(c)}{x-c}$$

exist and be the same. These lateral limits are called derived by the left and by the right, respectively. It is said that f is derivable in a closed interval [a, b] if it is derivable in (a, b) and there is also the derivative from the right in a and the derivative from the left in b.

If a function is not continuous at x = c, it cannot be derivable at x = c. For example, the integer part or greater whole function.

APPLICATION OF THE LIMIT PROCESS FOR THE CALCULATION OF DERIVATIVES

To find the derivative of a function requires the definition of the derivative of a function, that is   the derivative of f in x is given by

$$f'(x) = \lim_{\Delta x \to 0} \frac{f(x+\Delta x)-f(x)}{\Delta x}$$

Provided that limit exists.

Differentiation rule: The rule of constant

La derivada de una función constante es 0. Es decir, si c es un número real, entonces

$$\frac{d}{dx}[c] = 0$$

Differentiation rule: Power rule

If n is a rational number, then the function$(x) = x^n$ is derivable y $\frac{d}{dx}[x^n] = nx^{n-1}$.

The power rule simply states that to differentiate x^n: The exponent is placed in front of x^n the exponent is decreased by 1.

Differentiation rule: The constant multiple rule

If f is a derivable function and c a real number, then cf is derivable and:

$$\frac{d}{dx}[cf(x)] = cf'(x)$$

Differentiation rules: The rules of addition and difference

The derivative of the sum (or difference) of two derivable functions f and g is itself derivable. In addition, the derivative R + r (or f-g) is equal to the sum (or difference) of the derivatives of f, g.

$$\frac{d}{dx}[f(x) + g(x)] = f'(x) + g'(x)$$

$$\frac{d}{dx}[f(x) - g(x)] = f'(x) - g'(x)$$

DERIVED FROM THE SINE AND COSINE FUNCTIONS

$$\lim_{\Delta x \to 0} \frac{sen\, \Delta x}{\Delta x} = 1 \quad and \quad \lim_{\Delta x \to 0} \frac{1 - cos\, \Delta x}{\Delta x} = 0$$

These two limits can be used to demonstrate the derivation rules of the sine and cosine functions. And these in turn serve as a basis to demonstrate the remaining trigonometric functions.

Differentiation rule: Product rule

If f and g are differentiable functions, then

$$\frac{d}{dx}[f(x)\, g(x)] = f(x)\, g'(x) + g(x)\, f'(x)$$

The rule of the product is usually stored verbally in the following way: The first function by the derivative of the second function plus the second function by the derivative of the first.

Differentiation rule: Quotient rule

If f and g are differentiable functions and g (x) ≠ 0, then

$$\frac{d}{dx}\left[\frac{f(x)}{g(x)}\right] = \frac{g(x)f'(x) - f(x)g'(x)}{[g(x)]^2}$$

In verbal form, the quotient rule is: The denominator by the derivative of the numerator, minus the numerator multiplied by the derivative of the denominator, all divided by the denominator squared.

Differentiation rule: The rule of the chain

If Y = f (u) is a derivable function of u and besides u = g (x) is a derivable function of x, then Y = f (g (x)) is a derivable function of x.

$$\frac{dy}{dx} = \frac{dy}{du}\frac{du}{dx}$$

Or its equivalent

$$\frac{d}{dx}[f(g(x))] = f'(g(x)) * g'(x)$$

THE GENERAL RULE OF POWERS

The rule for deriving such functions is called the general rule of the powers, and it is but a particular case of the rule of the chain.

yes, $y = [u(x)]^n$, where u is a derivable function of x and n is a rational number, then

$$\frac{dy}{dx} = n[u(x)]^{n-1}\frac{du}{dx}$$

Or its equivalent

$$\frac{d}{dx}[u^n] = nu^{n-1}u'$$

IMPLICIT DERIVATION

A function y = f (x), where its dependent variable is expressed only in terms of the independent variable x, is known as an explicit function.

$$y = 2X^2 - 6$$

However, some functions are only implicitly stated in an equation. For example,

$$x^2 - 2y^3 + 4y = 2$$

Then the following question arises: How to find the $\dfrac{dy}{dx}$ for the previous equation where it is very difficult to clear and as an explicit function of x?

In these types of situations, the so-called implicit derivation must be used. To be clear about this technique it is necessary to take into account that the derivation is made with respect to x. If there is, a term to be derived where (and) it will be necessary to apply the chain rule.

For example, $\dfrac{d}{dx}[y^5] = 5y^4 \dfrac{dy}{dx}$

STRATEGIES FOR THE IMPLICIT DERIVATION

a) Derive both sides of the equation with respect to x.

b) Group all the terms in which $\dfrac{dy}{dx}$ appears on the left side of the equation and pass all the others to the right.

c) Factor $\dfrac{dy}{dx}$ from the left side of the equation.

d) Clear $\dfrac{dy}{dx}$.

PACE OR RELATED REASONS FOR CHANGE

The derivative $\dfrac{dy}{dx}$ of a function y = f (x) is its instantaneous rate of change with respect to the variable x. When a function describes position or distance, then its rate of change with respect to time is interpreted as velocity. In general, a rate of change (or intensity of variation) with respect to time is the answer to the question How fast does a quantity vary? For example, if V represents a volume that varies or changes over time, then $\dfrac{dV}{dt}$ is the ratio, or the rate at which the volume is changing at time t. One reason, for example, $\dfrac{dV}{dt} = 10\dfrac{cm^3}{s}$, means that the volume is increasing 10 cubic centimeters every second. Similarly, if a person is walking towards the lamppost, at a constant rate of 3 feet / s, then $\dfrac{dx}{dt} = -3$ feet / s. On the other hand, if the person walks away from the pole then $\dfrac{ax}{dt} = 3$ feet / s. The negative and positive reasons mean, of course, that the distance x is decreasing and growing, respectively.

STRATEGY TO SOLVE RHYTHM PROBLEMS OR RELATED REASONS
FOR CHANGE

1. If possible, draw a diagram that illustrates the situation.

2. Designate with symbols all the quantities given and the quantities to be determined that vary with time.

3. Analyze the statement of the problem and distinguish which reasons for change are known

4. and what is the reason or pace of change that is required.

5. Present an equation that relates the variables whose reasons for change are given or have to be determined.

6. Using the chain rule, implicitly derive both members of the equation obtained in the previous section, with respect to time.

7. Substitute in the equation resulting from point (5), all the known values change their reasons for change, in order to deduce (clear) the required rate of change.

Function optimization

In science, engineering and administration, it is common to be interested in the maximum and minimum values of functions; For example, a company is naturally interested in maximizing revenue while minimizing costs. The next time the reader goes to a supermarket, try this experiment: take a small ruler with you and measure the height and diameter of all the cans that contain, for example, 16 ounces of food (28.9 plg^3). The fact that all cans of this specified volume have the same measurements is not a coincidence, since there are specific dimensions that will minimize the amount of metal used and, therefore, minimize the manufacturing cost to the company. In the problems that follow, a function will be given, or else the verbal description will have to be interpreted to establish a function from which a maximum or minimum value is sought. These are the types of verbal problems that enhance the power of calculation and provide one of the many possible answers to the old question of: what is the use? Next, the important steps in the solution of a problem of maximum and minimum application are indicat.

1. Identify all the quantities given and those to be determined. If possible, draw a drawing.

2. Write a primary equation for the amount that will be maximized or minimized.

3. Reduce the primary equation to one that has a single independent variable. This may involve the use of secondary equations that relate the independent variables of the primary equation.

4. Determine the admissible domain of the primary equation. That is, determine the values for which the problem raised makes sense.

5. Determine the maximum or minimum value desired by calculation techniques.

DERIVATIVES OF A HIGHER ORDER

Just as deriving a position function, a velocity function is obtained, when deriving the latter, an acceleration function is obtained. In other words, the acceleration function is the second derivative of the position function. The second derivative is an example of a higher order derivative.

Ends of a function

It is of vital importance to determine the behavior of a function in an interval I. The following questions are pertinent: does f have a maximum value? Does it have a minimum value? Where is the function growing? Where is it decreasing? In this section of the theoretical foundation you will see how the derivatives are used to answer these questions.

End definition

Let f be a function defined in a range I containing a c.

f (c) is the minimum of f in I if f (c) ≤ f (x) for all x in I.

f (c) is the maximum of f in I if f (c) ≥ f (x) n for all x in I.

Extreme value theorem

If f is continuous in the closed interval [a, b], then f has a minimum as a maximum in the interval.

Relative extremes and critical points or numbers

Definition of relative extremes:

1. If there is an open interval containing ac in which f (c) is a maximum, then f (c) is called the relative maximum of f, or it could be stated that f has a relative maximum in (c, f (c)).

2. If there is an open interval containing ac in which f (c) is a minimum, then f (c) is called the relative minimum of f, or one could say that f has a relative minimum in (c, f) (c)).

Definition of a number or critical point:

Let f be defined in c. If f '(c) = 0 or if f is not derivable in c, then c is a critical point of f.

Theorem:

Relative extremes occur only in numbers or critical points.

It is concluded, if f has a relative minimum or a relative maximum at x = c, then c is a critical point of f.

Determination of ends in a closed interval

The previous theorem states that the relative ends of a function can only occur at the critical points of the function. The following strategies can be used to determine the extremes in a closed interval.

Strategies for determining ends in a closed interval:

1. The critical points of f are found (a, b).

2. f is evaluated at each critical point in (a, b).

3. f is evaluated at each endpoint of [a, b].

4. The smallest of these values is the minimum. The largest is the maximum.

Determine the extremes of $f(x) = 3x^4 - 4x^3$ in the interval [-1, 2].
By applying the above strategies you get the following results:
Critical points: 0 and 1, minimum = -1, maximum = 16.

Criterion of the second derivative

Theorem:
Let f be a function such that f'(c) = 0 and the second derivative of f exists in an open interval that contains a c.

1. If f'' (c)> 0, then f has a relative minimum in (c, f (c))

2. If f'' (c) <0, then f has a relative maximum in (c, f (c))

If f'' = 0, then the criterion fails.

TABLES AND MATHEMATICAL FORMULAS

PERIMETERS, AREAS AND VOLUMES

Triangle 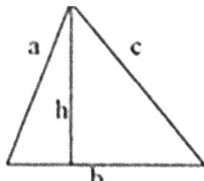 $p = a + b + c$

$$A = \frac{b \cdot h}{2}$$

Rectangle 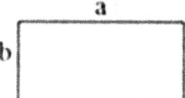 $p = 2a + 2b$

$$A = a \cdot b$$

Hexagon 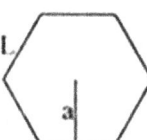 $p = 6L$

$$A = \frac{p \cdot a}{2}$$

Circle 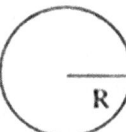 $\text{Cfe. Leng} = 2\pi R$

$$A = \pi R^2$$

Circular sector 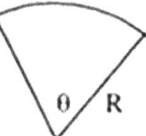 $\text{Arc. Leng} = R\theta$

$$A = \frac{1}{2}R^2\theta$$

Sphere   **Cylinder**   **Cone**

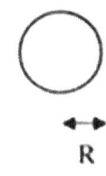   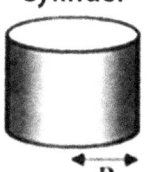   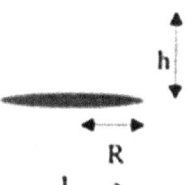

$A = 4\pi R^2$ $A_{total} = 2\pi R^2 + 2\pi Rh$ $V = \frac{1}{3}\pi R^2 h$

$V = \frac{4}{3}\pi R^3$ $V = \pi R^2 h$

TRIGONOMETRY

Units of measure of angles

$$\begin{cases} Degrees \\ Radians \end{cases}$$

Equivalence: $360^0 = 2\pi$ rad. $\Longrightarrow$ 1 rad $= \dfrac{180^0}{\pi} \cong 57^0 \, 17^m$

Length of an arc of a circle of radius R that subtends a central angle θ

$$s = R\theta \qquad \theta \quad \text{in radians}$$

Values of trigonometric lines of some remarkable angles

θ Degrees	0	30	45	60	90	120	180	270	360
θ Radian	0	$\dfrac{\pi}{6}$	$\dfrac{\pi}{4}$	$\dfrac{\pi}{3}$	$\dfrac{\pi}{2}$	$\dfrac{2\pi}{3}$	π	$\dfrac{3\pi}{2}$	2π
sen θ	0	$\dfrac{1}{2}$	$\dfrac{\sqrt2}{2}$	$\dfrac{\sqrt3}{2}$	1	$\dfrac{\sqrt3}{2}$	0	-1	0
cos θ	1	$\dfrac{\sqrt3}{2}$	$\dfrac{\sqrt2}{2}$	$\dfrac{1}{2}$	0	$-\dfrac{1}{2}$	-1	0	1
tg θ	0	$\dfrac{\sqrt3}{3}$	1	$\sqrt3$	$\exists$	$-\sqrt3$	0	$\exists$	0

Supplementary angles $\theta + \varphi = \pi$

sen θ = sen $(\pi-\theta)$ cos θ = - cos $(\pi-\theta)$ tg θ = - tg $(\pi-\theta)$

Complementary angles $\theta + \varphi = \dfrac{\pi}{2}$

sen θ = cos $(\dfrac{\pi}{2} - \theta)$ tg θ = cotg $(\dfrac{\pi}{2} - \theta)$

Opposite angles

Sen $(-\theta)$ = - sen θ cos $(-\theta)$ = cos θ tg $(-\theta)$ = - tg θ

Angles that differ in $\dfrac{\pi}{2}$ And in π

$$\text{sen}\left(\theta+\dfrac{\pi}{2}\right)=\cos\theta \qquad \cos\left(\theta+\dfrac{\pi}{2}\right)=-\text{sen}\,\theta \qquad \text{tg}\left(\theta+\dfrac{\pi}{2}\right)=-\cot g\,\theta$$

$$\text{sen}(\theta+\pi)=-\text{sen}\,\theta \qquad \cos(\theta+\pi)=-\cos\theta \qquad \text{tg}(\theta+\pi)=\text{tg}\,\theta$$

Sine theorem

$$\dfrac{\text{sen}A}{a}=\dfrac{\text{sen}B}{b}=\dfrac{\text{sen}C}{c}$$

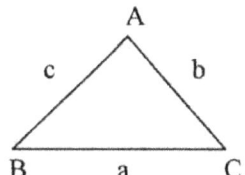

Cosine theorem

$$a^2=b^2+c^2-2\,b\,c\,\cos A$$

$$b^2=a^2+c^2-2\,a\,c\,\cos B$$

$$c^2=a^2+b^2-2\,a\,b\,\cos C$$

Fundamental formula

$$\text{sen}^2x+\cos^2x=1$$

Formula of addition and subtraction of angles

$$\text{sen}(x-y)=\text{sen}\,x\cos y-\cos x\,\text{sen}\,y$$

$$\cos(x+y)=\cos x\cos y-\text{sen}\,x\,\text{sen}\,y$$

$$\cos(x-y)=\cos x\cos y+\text{sen}\,x\,\text{sen}\,y$$

$$\text{tg}(x+y)=\dfrac{\text{tg}x+\text{tg}y}{1-\text{tg}x\,\text{tg}y}$$

$$\text{tg}(x-y)=\dfrac{\text{tg}x-\text{tg}y}{1+\text{tg}x\,\text{tg}y}$$

Formula of the double

$$\text{sen}\,2x=2\,\text{sen}x\cos x \qquad \cos 2x=\cos^2x-\text{sen}^2x \qquad \text{tg}\,2x=\dfrac{2\text{tg}x}{1-\text{tg}^2x}$$

Formula of the angle half

$$\text{sen}^2x=\dfrac{1-\cos 2x}{2} \qquad\qquad \cos^2x=\dfrac{1+\cos 2x}{2}$$

DERIVATIVE TABLE

$f(x)$	$\dfrac{df}{dx}$	$f(x)$	$\dfrac{df}{dx}$
k	0	senx	cosx
x	1	cosx	$-$ sen x
$\|x\|$	$sg(x) \quad x \neq 0$	tgx	$1 + tg^2 x$
x^m	mx^{m-1}	Arcsenx	$\dfrac{1}{\sqrt{1-x^2}}$
$\dfrac{1}{x}$	$-\dfrac{1}{x^2}$	Arccosx	$\dfrac{1}{-\sqrt{1-x^2}}$
$\sqrt{x}$	$\dfrac{1}{2\sqrt{x}}$	Arctgx	$\dfrac{1}{1+x^2}$
$\sqrt[3]{x}$	$\dfrac{1}{3\sqrt[3]{x^2}}$	shx	chx
e^x	e^x	chx	shx
Lx	$\dfrac{1}{x}$	thx	$1 - th^2 x$
$L\|x\|$	$\dfrac{1}{x}$	Argshx	$\dfrac{1}{\sqrt{x^2+1}}$
$Sg(x)$	$0 \quad \forall x \neq 0$	Argchx	$\dfrac{1}{\sqrt{x^2-1}}$
a^x	$a^x La$	Argthx	$\dfrac{1}{1-x^2}$

DERIVATIVES OF COMPOSITE FUNCTIONS

$(f \circ g)(x)$	$\dfrac{d(f \circ g)}{dx}$	$(f \circ g)(x)$	$\dfrac{d(f \circ g)}{dx}$
$g(x)$	$\dfrac{dg}{dx}$	$\operatorname{sen} g(x)$	$\cos g. \dfrac{dg}{dx}$
$k.g$	$k\dfrac{dg}{dx}$	$\cos g(x)$	$-\operatorname{sen} g. \dfrac{dg}{dx}$
$\lvert g \rvert$	$sg(g). \dfrac{dg}{dx}$	$\operatorname{tg} g(x)$	$(1 + \operatorname{tg}^2 g). \dfrac{dg}{dx}$
g^m	$mg^{m-1}\dfrac{dg}{dx}$	$\operatorname{Arcsen} g(x)$	$\dfrac{1}{\sqrt{1-x^2}} \dfrac{dg}{dx}$
$\dfrac{1}{g}$	$-\dfrac{1}{g^2}\dfrac{dg}{dx}$	$\operatorname{Arccos} g(x)$	$-\dfrac{1}{\sqrt{1-x^2}} \dfrac{dg}{dx}$
$\sqrt{g}$	$\dfrac{1}{2\sqrt{g}}\dfrac{dg}{dx}$	$\operatorname{Arctg} g(x)$	$\dfrac{1}{1+g^2} \dfrac{dg}{dx}$
$\sqrt[3]{g}$	$\dfrac{1}{3\sqrt[3]{g^2}}\dfrac{dg}{dx}$	$\operatorname{sh} g(x)$	$\operatorname{ch} g(x). \dfrac{dg}{dx}$
e^g	$e^g\dfrac{dg}{dx}$	$\operatorname{ch} g(x)$	$\operatorname{sh} g(x). \dfrac{dg}{dx}$
$Lg \quad o \; L\lvert g\rvert$	$\dfrac{1}{g}\dfrac{dg}{dx}$	$\operatorname{th} g(x)$	$(1 - \operatorname{th}^2 g)\dfrac{dg}{dx}$
$L\left\lvert\dfrac{g}{h}\right\rvert$	$\dfrac{1}{g}\dfrac{dg}{dx} - \dfrac{1}{h}\dfrac{dh}{dx}$	$\operatorname{Argsh} g(x)$	$\dfrac{1}{\sqrt{1+g^2}} \dfrac{dg}{dx}$
a^g	$a^g.La. \dfrac{dg}{dx}$	$\operatorname{Argch} g(x)$	$\dfrac{1}{\sqrt{g^2-1}} \dfrac{dg}{dx}$
g^h	$g^h\left[\dfrac{dh}{dx}Lg + \dfrac{h}{g}\dfrac{dg}{dx}\right]$	$\operatorname{Argth} g(x)$	$\dfrac{1}{1-g^2} \dfrac{dg}{dx}$
$h\,e^g$		$e^g\left[\dfrac{dh}{dx} + h.\dfrac{dg}{dx}\right]$	

BIBLIOGRAPHY

Larson R, Hostetler R. Y Edwards B. (1995).
Calculation. Volume 1. México. McGraw-Hill.

Zill G, Dennis (1985).
Calculation with Analytical Geometry. Mexico, Iberoamerica Editorial Group.

Wisniewski Piotr M. y Banegas G. Ana L. (2004).
Introduction to university mathematics. Mexico. McGraw-Hill.

Pita R, Claudio. (1998).
Calculation of a variable. Mexico, Prentice- Hispano-American Hall.

CALCULATION OF DERIVATIVES APPLYING THE DEFINITION OF LIMITS

$$f(x) = -5x$$

001

To find the derivative of a function, it is necessary to define the derivative of a function, that is: The derivative of f in x is given by

$$f'(x) = \lim_{\Delta x \to 0} \frac{f(x+\Delta x)-f(x)}{\Delta x}$$

Provided that limit exists

The expression $x + \Delta x$ is substituted in the variable x of the function

$$f'(x) = \lim_{\Delta x \to 0} \frac{-5(x+\Delta x)-(-5x)}{\Delta x}$$

Working mathematically to eliminate parentheses, you have

$$f'(x) = \lim_{\Delta x \to 0} \frac{-5x-5\Delta x+5x}{\Delta x}$$

Similar terms are canceled

$$f'(x) = \lim_{\Delta x \to 0} \frac{-5\Delta x}{\Delta x}$$

The Δx are canceled

$$f'(x) = \lim_{\Delta x \to 0}(-5)$$

The limit of a constant is the constant

$$f'(x) = -5$$

002

$$h(s) = 3 + \frac{2}{3}s$$

To find the derivative of a function, it is necessary to define the derivative of a function, that is: The derivative of f in x is given by

$$f'(x) = \lim_{\Delta x \to 0} \frac{f(x+\Delta x)-f(x)}{\Delta x}$$

Provided that limit exists. Note that the independent variable is S

$$h'(s) = \lim_{\Delta s \to 0} \frac{h(s+\Delta s)-h(s)}{\Delta s}$$

The expression $s + \Delta s$ is substituted in the variable s of the function

$$h'(s) = \lim_{\Delta s \to 0} \frac{3+\frac{2}{3}(s+\Delta s)-(3+\frac{2}{3}s)}{\Delta s}$$

The indicated operations are performed to eliminate the parentheses

$$h'(s) = \lim_{\Delta s \to 0} \frac{3+\frac{2}{3}s+\frac{2}{3}\Delta s-3-\frac{2}{3}s}{\Delta s}$$

Similar terms are canceled

$$h'(s) = \lim_{\Delta s \to 0} \frac{\frac{2}{3}\Delta s}{\Delta s} \quad \rightarrow \quad h'(s) = \lim_{\Delta s \to 0} \frac{2}{3}$$

The limit of a constant is the constant

$$\boxed{h'(s) = \frac{2}{3}}$$

$$f(x) = 2x^2 + x - 1$$

003

To find the derivative of a function, it is necessary to define the derivative of a function, that is: The derivative of f in x is given by

$$f'(x) = \lim_{\Delta x \to 0} \frac{f(x+\Delta x) - f(x)}{\Delta x} \quad \to \quad \text{Provided that limit exists}$$

The expression $x + \Delta x$ is substituted in the variable x of the function

$$f'(x) = \lim_{\Delta x \to 0} \frac{[2(x+\Delta x)^2 + (x+\Delta x) - 1] - (2x^2 + x - 1)}{\Delta x}$$

Solve the square of the sum and operate mathematically

$$f'(x) = \lim_{\Delta x \to 0} \frac{\left[2(x^2 + 2x\Delta x + (\Delta x)^2) + x + \Delta x - 1\right] - 2x^2 - x + 1}{\Delta x}$$

$$f'(x) = \lim_{\Delta x \to 0} \frac{2x^2 + 4x\Delta x + 2(\Delta x)^2 + x + \Delta x - 1 - 2x^2 - x + 1}{\Delta x}$$

$$f'(x) = \lim_{\Delta x \to 0} \frac{2(\Delta x)^2 + 4x\Delta x + \Delta x}{\Delta x}$$

$$f'(x) = \lim_{\Delta x \to 0} \frac{\Delta x(2\Delta x + 4x + 1)}{\Delta x} \quad \leftarrow \quad Common\ factor\ \Delta x\ is\ extracted$$

$$f'(x) = \lim_{\Delta x \to 0} 2\Delta x + 4x + 1$$

Evaluating the resulting limit, you get $\quad f'(x) = \lim_{\Delta x \to 0} 2\Delta x + 4x + 1 = 4x + 1$

Therefore, the derivative of $f(x) = 2x^2 + x - 1$ is: $4x + 1$

$$\boxed{f'(x) = 4x + 1}$$

004

$$f(x) = 3$$

The derivative of f in x is given by

$$f'(x) = \lim_{\Delta x \to 0} \frac{f(x+\Delta x)-f(x)}{\Delta x} \rightarrow \text{Provided that limit exists}$$

It is obvious that f $(x + \Delta x)$ is equal to 3, therefore

$$f'(x) = \lim_{\Delta x \to 0} \frac{3-3}{\Delta x} \quad \rightarrow \quad f'(x) = \lim_{\Delta x \to 0} \frac{0}{\Delta x} \quad \rightarrow$$

$$\rightarrow \quad f'(x) = \lim_{\Delta x \to 0} (0)$$

The limit of a constant is the constant, consequently $\boxed{f'(x) = 0}$

In conclusion, we have that the derivative of a constant is zero

005

$$f(x) = -5$$

The derivative of f in x is given by

$$f'(x) = \lim_{\Delta x \to 0} \frac{f(x+\Delta x)-f(x)}{\Delta x} \rightarrow \text{Provided that limit exists}$$

It is obvious that f $(x + \Delta x)$ is equal to -5, therefore

$$f'(x) = \lim_{\Delta x \to 0} \frac{-5-(-5)}{\Delta x} \quad \rightarrow \quad f'(x) = \lim_{\Delta x \to 0} \frac{0}{\Delta x} \quad \rightarrow$$

$$\rightarrow \quad f'(x) = \lim_{\Delta x \to 0} \frac{0}{\Delta x} = f'(x) = \lim_{\Delta x \to 0} 0$$

The limit of a constant is the constant, consequently

$$\boxed{f'(x) = 0}$$

$$f(x) = 1 - x^2$$

006

The derivative of f in x is given by

$$f'(x) = \lim_{\Delta x \to 0} \frac{f(x+\Delta x) - f(x)}{\Delta x} \rightarrow \text{ Provided that limit exists}$$

The expression $x + \Delta x$ is substituted in the variable x of the function and is operated mathematically

$$f'(x) = \lim_{\Delta x \to 0} \frac{\left[1-(x+\Delta x)^2\right]-(1-x^2)}{\Delta x}$$

$$f'(x) = \lim_{\Delta x \to 0} \frac{\left[1-(x^2+2x\Delta x+(\Delta x)^2)\right]-1+x^2}{\Delta x}$$

$$f'(x) = \lim_{\Delta x \to 0} \frac{1-x^2-2x\Delta x-(\Delta x)^2-1+x^2}{\Delta x}$$

$$f'(x) = \lim_{\Delta x \to 0} \frac{-(\Delta x)^2-2x\Delta x}{\Delta x}$$

$$f'(x) = \lim_{\Delta x \to 0} \frac{-\Delta x(\Delta x+2x)}{\Delta x} \quad \leftarrow \text{ } Common \text{ } factor \text{ } \Delta x \text{ } extracted$$

$$f'(x) = \lim_{\Delta x \to 0} -(\Delta x + 2x)$$

Evaluating the resulting limit

$$f'(x) = \lim_{\Delta x \to 0} -(\Delta x + 2x) = -2x$$

Consequently, we have that the derivative of the function is

$$\boxed{f'(x) = -2x}$$

007

Find the slopes of the tangent lines of the following function

$$f(x) = x^2 + 1 \text{ In points } (0,1) \text{ y } (-1,2)$$

Let $(c, f(c))$ be any point of the graph of f. The slope (m) of the tangent line at the point is found by

$$m = \lim_{\Delta x \to 0} \frac{f(c+\Delta x)-f(c)}{\Delta x}$$

Replace the expression $c + \Delta x$ and c in the variable of x of the function

$$m = \lim_{\Delta x \to 0} \frac{[(c+\Delta x)^2+1]-(c^2+1)}{\Delta x} \quad \to \quad m = \lim_{\Delta x \to 0} \frac{c^2+2c\Delta x+(\Delta x)^2+1-c^2-1}{\Delta x} \quad \to$$

$$m = \lim_{\Delta x \to 0} \frac{2c\Delta x+(\Delta x)^2}{\Delta x} \quad \to \quad m = \lim_{\Delta x \to 0} \frac{\Delta x(2c+\Delta x)}{\Delta x} \quad \to$$

$$m = \lim_{\Delta x \to 0} \frac{\Delta x(2c+\Delta x)}{\Delta x} \quad \to \quad m = \lim_{\Delta x \to 0} (2c + \Delta x)$$

Evaluating the resulting limit

$$m = \lim_{\Delta x \to 0} (2c + \Delta x) = 2c$$

So, **m = 2c**

On the point (0, 1), the slope is **m = 2(0) = 0**

On the point (-1, 2), the slope is **m = 2(-1) = −2**

Prove that the derivative of the sine of *x* is the cosine of *x*. That is to say:
$$\frac{d}{dx}[sen\,x] = \cos x.$$

008

Demonstration

$$\frac{d}{dx}[sen\,x] = \lim_{\Delta x \to 0} \frac{sen\,(x+\Delta x) - sen\,x}{\Delta x}$$

By applying the sine of the sum of two angles

$$\frac{d}{dx}[sen\,x] = \lim_{\Delta x \to 0} \frac{sen\,x\cos\Delta x + sen\,\Delta x\cos x - sen\,x}{\Delta x}$$

$$\frac{d}{dx}[sen\,x] = \lim_{\Delta x \to 0} \frac{\cos x\,sen\,\Delta x - sen\,x(1-\cos\Delta x)}{\Delta x}$$

$$\frac{d}{dx}[sen\,x] = \lim_{\Delta x \to 0}\cos x\left(\frac{sen\,\Delta x}{\Delta x}\right) - \lim_{\Delta x \to 0}sen\,x\left(\frac{1-\cos\Delta x}{\Delta x}\right)$$

$$\frac{d}{dx}[sen\,x] = \cos x\left(\lim_{\Delta x \to 0}\frac{sen\,\Delta x}{\Delta x}\right) - sen\,x\left(\lim_{\Delta x \to 0}\frac{(1-\cos\Delta x)}{\Delta x}\right)$$

As $\qquad \lim_{\Delta x \to 0}\dfrac{sen\,\Delta x}{\Delta x} = 1 \qquad and \qquad \lim_{\Delta x \to 0}\dfrac{1-\cos\Delta x}{\Delta x} = 0$

In consecuense

$$\frac{d}{dx}[sen\,x] = \cos x(1) - sen\,x(0) = \cos x$$

With this procedure it is demonstrated: $\dfrac{d}{dx}[sen\,x] = \cos x.$

009

Show that the cosine derivative of x is minus the sine of x. That is to say:

$$\frac{d}{dx}\left[\cos x\right] = -sen\, x$$

Demonstration

$$\frac{d}{dx}\left[\cos x\right] = \lim_{\Delta x \to 0} \frac{\cos(x+\Delta x) - \cos x}{\Delta x}$$

When applying the cosine of the sum of two angles

$$\frac{d}{dx}\left[\cos x\right] = \lim_{\Delta x \to 0} \frac{\cos x \cos \Delta x - sen\, x\, sen\, \Delta x - \cos x}{\Delta x}$$

$$\frac{d}{dx}\left[\cos x\right] = \lim_{\Delta x \to 0} \frac{-\cos x(1-\cos \Delta x) - sen\, x\, sen\, \Delta x}{\Delta x}$$

$$\frac{d}{dx}\left[\cos x\right] = \lim_{\Delta x \to 0}\left[-\cos x\left(\frac{1-\cos \Delta x}{\Delta x}\right) - sen\, x\left(\frac{sen\, \Delta x}{\Delta x}\right)\right]$$

$$\frac{d}{dx}\left[\cos x\right] = -\cos x \lim_{\Delta x \to 0}\left(\frac{1-\cos \Delta x}{\Delta x}\right) - sen\, x \lim_{\Delta x \to 0}\left(\frac{sen\, \Delta x}{\Delta x}\right)$$

As $\qquad \lim_{\Delta x \to 0} \frac{sen\, \Delta x}{\Delta x} = 1 \qquad and \qquad \lim_{\Delta x \to 0} \frac{1-\cos \Delta x}{\Delta x} = 0$

In consecuense

$$\frac{d}{dx}\left[\cos x\right] = -\cos x\,(0) - sen\, x\,(1) = -sen\, x$$

With this procedure it is demonstrated: $\frac{d}{dx}\left[\cos x\right] = -sen\, x.$

$$f(x) = 3x + 2$$

010

The derivative of f in x is given by

$$f'(x) = \lim_{\Delta x \to 0} \frac{f(x+\Delta x) - f(x)}{\Delta x} \rightarrow \text{ Provided that limit exists}$$

The expression $x + \Delta x$ is substituted in the variable x of the function

$$f'(x) = \lim_{\Delta x \to 0} \frac{[3(x+\Delta x)+2] - (3x+2)}{\Delta x}$$

It operates mathematically to eliminate the parentheses

$$f'(x) = \lim_{\Delta x \to 0} \frac{[3x+3\Delta x+2] - 3x - 2}{\Delta x}$$

Removed the bracket

$$f'(x) = \lim_{\Delta x \to 0} \frac{3x+3\Delta x+2-3x-2}{\Delta x}$$

Operating to cancel similar terms

$$f'(x) = \lim_{\Delta x \to 0} \frac{3\Delta x}{\Delta x} \rightarrow f'(x) = \lim_{\Delta x \to 0} 3$$

Evaluating the limit

$$f'(x) = \lim_{\Delta x \to 0} 3 = 3$$

The limit of a constant is the constant, therefore, the derivative of the function is:

$$\boxed{f'(x) = 3}$$

DIFFERENTIATION RULE: Power rule

If n is a rational number, then the function f (x) = x^n is derivable and $\frac{d}{dx}[x^n] = nx^{n-1}$. The rule of power simply states that to differentiate x^n the exponent is placed before x^n and one decreases the exponent.

 011 $f(x) = x^3$

The power rule is applied

$$\frac{d}{dx}[f(x)] = \frac{d}{dx}[x^3]$$
$$= 3x^{3-1}$$
$$= 3x^2$$

$$\boxed{\frac{d}{dx}[x^3] = 3x^2}$$

012 $g(x) = \sqrt[3]{x}$

Before applying the power rule, the function is rewritten in such a way that a potential function appears, that is

$$g(x) = \sqrt[3]{x} = x^{\frac{1}{3}} \;\to\; g(x) = x^{\frac{1}{3}}$$

$$\frac{d}{dx}[g(x)] = \frac{d}{dx}\left[x^{\frac{1}{3}}\right] = \frac{1}{3} x^{-\frac{2}{3}}$$

$$\frac{d}{dx}[g(x)] = \frac{1}{3x^{\frac{2}{3}}} \qquad \boxed{\frac{d}{dx}\left[\sqrt[3]{x}\right] = \frac{1}{3\sqrt[3]{x^2}}}$$

$$y = \frac{1}{x^2}$$

013

Before applying the power rule, the function is rewritten in such a way that a potential function appears, that is

$$y = \frac{1}{x^2} = x^{-2} \rightarrow y = x^{-2}$$

$$\frac{d}{dx}[y] = \frac{d}{dx}[x^{-2}]$$

$$= -2x^{-2-1}$$

$$= -2x^{-3}$$

$$= \frac{-2}{x^3} \qquad \boxed{\frac{d}{dx}\left[\frac{1}{x^2}\right] = -\frac{2}{x^3}}$$

$$y = 5x^3$$

014

Before applying the power rule, the function is rewritten by applying the rule of constant multiple. That is to say

$$\frac{d}{dx}[y] = \frac{d}{dx}[5x^3]$$

Rule of the constant multiple $\rightarrow$ $= 5\frac{d}{dx}[x^3] = 5(3x^2)$

$$= 15x^2$$

$$\boxed{\frac{d}{dx}[5x^3] = 15x^2}$$

Note: If C is any constant and f is differentiable, then

$$\frac{d}{dx}[C\,f(x)] = C\frac{d}{dx}f(x) \rightarrow \text{Rule of the constant multiple of a function.}$$

015

$$y = 4x^{12}$$

$$\frac{d}{dx}[y] = \frac{d}{dx}[4x^{12}] = 4\frac{d}{dx}[x^{12}]$$
$$= 4(12x^{11})$$
$$= 48x^{11}$$

$$\boxed{\frac{d}{dx}[4x^{12}] = 48x^{11}}$$

016

$$y = \frac{\pi^6}{12}$$

$$\frac{d}{dx}[y] = \frac{d}{dx}\left[\frac{\pi^6}{12}\right] = 0$$

$$\boxed{\frac{d}{dx}\left[\frac{\pi^6}{12}\right] = 0}$$

Note: The derivative of a constant is zero. See problem number 4.

017

$$y = x$$

$y = x^n$ for n = 1 you have

$$\frac{d}{dx}[y] = \frac{d}{dx}[x]$$

$$= \frac{dx}{dx} = 1 \quad \boxed{y' = 1}$$

If n = 1 then the derivative of x is 1.

$$y = \frac{1}{2\sqrt[3]{x^2}}$$

018

Before applying the power rule, the function is rewritten by applying the rule of constant multiple. That is to say

$$y = \frac{1}{2\sqrt[3]{x^2}} = \frac{1}{2x^{\frac{2}{3}}} = \frac{1}{2}\,x^{-\frac{2}{3}} \quad \rightarrow \quad y = \frac{1}{2}\,x^{-\frac{2}{3}}$$

$$\frac{dy}{dx} = -\frac{2}{6}\,x^{-\frac{2}{3}-1}$$

$$\frac{dy}{dx} = -\frac{1}{3}\,x^{-\frac{5}{3}} \qquad \boxed{\frac{dy}{dx} = -\frac{1}{3\sqrt[3]{x^5}}}$$

$$y = -\frac{3x}{2}$$

019

$$\frac{dy}{dx} = -\frac{3}{2}\frac{dx}{dx} \quad \rightarrow \quad \boxed{\frac{dy}{dx} = -\frac{3}{2}}$$

$$y = \frac{2}{x}$$

020

Before applying the power rule, the function is rewritten by applying the rule of constant multiple. That is to say

$$y = \frac{2}{x} = 2x^{-1} \rightarrow \frac{dy}{dx} = -2x^{-1-1} \rightarrow \boxed{\frac{dy}{dx} = -\frac{2}{x^2}}$$

47

. . . .
RULE OF DIFFERENTIATION: Rule of the sum and difference
. . . .

Note that each term in the function represents a differentiation rule. That is, the first term refers to the power rule. The second to the rule of the constant multiple of a function. And the third term to the derivative of a constant.

021

$$f(x) = x^3 - 4x + 5$$

$$\frac{d}{dx}[x^3 - 4x + 5] = \frac{d}{dx}(x^3) - 4\frac{d}{dx}(x) + \frac{d}{dx}(5)$$

$$= 3x^2 - 4 + 0$$
$$= 3x^2 - 4$$

$$\boxed{f'(x) = 3x^2 - 4}$$

022

$$g(x) = -\frac{x^4}{2} + 3x^3 - 2x$$

$$g(x) = -\frac{1}{2}x^4 + 3x^3 - 2x$$

$$g'(x) = -\frac{1}{2}\frac{d}{dx}(x^4) + 3\frac{d}{dx}(x^3) - 2\frac{d}{dx}(x)$$

$$= -\frac{4}{2}x^3 + 9x^2 - 2$$
$$= -2x^3 + 9x^2 - 2$$

$$\boxed{g'(x) = -2x^3 + 9x^2 - 2}$$

Note: The reader must bear in mind the importance of rewriting the proposed exercise.

$$y = 4x^4 - \frac{1}{x^2}$$

023

The function is rewritten before derivation

$$y = 4x^4 - 1(x^{-2}) \rightarrow y = 4x^4 - x^{-2}$$

$$y\,' = 4\frac{d}{dx}(x^4) - \frac{d}{dx}(x^{-2})$$

$$= 16x^3 + 2x^{-3}$$

$$= 16x^3 + \frac{2}{x^3}$$

$$\boxed{y' = 16x^3 + \frac{2}{x^3}}$$

$$y = \frac{x}{3} - \left(\frac{x}{4}\right)^2$$

024

The function is rewritten before derivation

$$y = \frac{x}{3} - \frac{x^2}{16} \rightarrow y = \frac{1}{3}x - \frac{1}{16}x^2$$

$$y' = \frac{1}{3}\frac{d}{dx}(x) - \frac{1}{16}\frac{d}{dx}(x^2)$$

$$y' = \frac{1}{3} - \frac{2}{16}x = -\frac{1}{8}x + \frac{1}{3}$$

$$\boxed{y' = -\frac{1}{8}x + \frac{1}{3}}$$

025

$$y = \frac{1}{2x} - \frac{1}{3x^2}$$

The function is rewritten before derivation

$$y = \frac{1}{2}x^{-1} - \frac{1}{3}x^{-2}$$

$$y' = \frac{1}{2}\frac{d}{dx}(x^{-1}) - \frac{1}{3}\frac{d}{dx}(x^{-2})$$

$$= -\frac{1}{2}x^{-2} + \frac{2}{3}x^{-3}$$

$$\boxed{y' = -\frac{1}{2x^2} + \frac{2}{3x^3}}$$

026

$$y = \frac{2x^3 - 3x^2 + 4x - 5}{x^2}$$

The function is rewritten before derivation

$$y = \frac{2x^3}{x^2} - \frac{3x^2}{x^2} + \frac{4x}{x^2} - \frac{5}{x^2} \;\rightarrow\; y = 2x - 3 + \frac{4}{x} - \frac{5}{x^2}$$

$$y = -5x^{-2} + 4x^{-1} + 2x - 3$$

$$y' = -5\frac{d}{dx}(x^{-2}) + 4\frac{d}{dx}(x^{-1}) + 2\frac{d}{dx}(x) - \frac{d}{dx}(3)$$

$$y' = 10x^{-3} - 4x^{-2} + 2$$

$$\boxed{y' = \frac{10}{x^3} - \frac{4}{x^2} + 2}$$

$$y = x^2 \left(2x^3 - \frac{x}{4x^4}\right)$$

027

Before deriving, it is operated algebraically to eliminate the parenthesis

$$y = 2x^5 - \frac{x^3}{4x^4} \quad \rightarrow \quad y = 2x^5 - \frac{1}{4x} \quad \rightarrow \quad y = 2x^5 - \frac{1}{4}x^{-1}$$

$$y' = 2\frac{d}{dx}(x^5) - \frac{1}{4}\frac{d}{dx}(x^{-1}) \quad \rightarrow \quad y' = 10x^4 + \frac{x^{-2}}{4}$$

$$y' = 10x^4 + \frac{1}{4x^2}$$

$$y = \frac{\pi}{2} sen\, \theta - \cos \theta$$

028

$$y' = \frac{\pi}{2}\frac{d}{d\theta}(sen\, \theta) - \frac{d}{d\theta}(\cos \theta)$$

$$y' = \frac{\pi}{2}\cos \theta - (-sen\, \theta)$$

$$y' = \frac{\pi}{2}\cos \theta + sen\, \theta$$

$$f(x) = x^2 + 5 - 3x^{-2}$$

029

$$f'(x) = \frac{d}{dx}(x^2) + \frac{d}{dx}(5) - 3\frac{d}{dx}(x^{-2})$$

$$f'(x) = 2x + 0 + 6x^{-3}$$

$$f'(x) = 2x + \frac{6}{x^3}$$

030

$$g(t) = t^2 - \frac{4}{t^3}$$

The function is rewritten before derivation

$$g(t) = t^2 - 4t^{-3}$$

$$g'(t) = \frac{d}{dx}(t^2) - 4\frac{d}{dt}(t^{-3})$$

$$g'(t) = 2t + 12t^{-4}$$

$$\boxed{g'(t) = 2t + \frac{12}{t^4}}$$

031

$$f(x) = \frac{x^3 - 3x^2 + 4}{x^2}$$

The function is rewritten before derivation

$$f(x) = \frac{x^3}{x^2} - \frac{3x^2}{x^2} + \frac{4}{x^2} \rightarrow f(x) = x - 3 + 4x^{-2}$$

$$f'(x) = \frac{d}{dx}(x) - \frac{d}{dx}(3) + 4\frac{d}{dx}(x^{-2})$$

$$f'(x) = 1 - 0 - 8x^{-3}$$

$$f'(x) = 1 - \frac{8}{x^3}$$

$$\boxed{f'(x) = \frac{x^3 - 8}{x^3}}$$

$$y = x(x^2 + 1)$$

032

The function is rewritten

$$y = x^3 + x$$

$$y' = \frac{d}{dx}(x^3) + \frac{d}{dx}(x) \quad \boxed{\boldsymbol{y' = 3x^2 + 1}}$$

$$f(x) = \sqrt{x} - 6\sqrt[3]{x}$$

033

The function is rewritten before derivation

$$f(x) = x^{\frac{1}{2}} - 6x^{\frac{1}{3}}$$

$$f'(x) = \frac{d}{dx}\left(x^{\frac{1}{2}}\right) - 6\frac{d}{dx}\left(x^{\frac{1}{3}}\right)$$

$$f'(x) = \frac{1}{2}x^{-\frac{1}{2}} - \frac{6}{3}x^{-\frac{2}{3}} \quad \boxed{\boldsymbol{f'(x) = \frac{1}{2\sqrt{x}} - \frac{2}{\sqrt[3]{x^2}}}}$$

$$h(s) = s^{\frac{4}{5}} - s^{\frac{2}{3}}$$

034

$$h'(s) = \frac{4}{5}s^{-\frac{1}{5}} - \frac{2}{3}s^{-\frac{1}{3}} \quad \boxed{\boldsymbol{h'(s) = \frac{4}{5s^{\frac{1}{5}}} - \frac{2}{3s^{\frac{1}{3}}}}}$$

035

$$f(x) = 6\sqrt{x} + 5\cos x$$

The function is rewritten before derivation

$$f(x) = 6x^{\frac{1}{2}} + 5\cos x$$

$$f'(x) = \frac{6}{2}x^{-\frac{1}{2}} + 5(-sen\ x)$$

$$f'(x) = \frac{6}{2}x^{-\frac{1}{2}} - 5senx$$

$$f'(x) = \frac{3}{\sqrt{x}} - 5sen\ x$$

036

$$y = \frac{5}{(2x)^3} + 2\cos x$$

We proceed to rewrite the function by first resolving the power

$$y = \frac{5}{(2x)^3} + 2\cos x \quad \rightarrow \quad y = \frac{5}{8x^3} + 2\cos x$$

$$y = \frac{5}{8}x^{-3} + 2\cos x \quad \rightarrow \quad y' = -\frac{15}{8}x^{-4} - 2sen\ x$$

$$y' = -\frac{15}{8}x^{-4} - 2sen\ x$$

$$f(x) = \frac{2}{\sqrt[3]{x}} + 3\cos x$$

037

The function is rewritten before derivation

$$f(x) = \frac{2}{\sqrt[3]{x}} + 3\cos x \quad \rightarrow \quad f(x) = 2x^{-\frac{1}{3}} + 3\cos x$$

$$f'(x) = -\frac{2}{3}\, x^{-\frac{4}{3}} + 3(-sen\, x)$$

$$\boxed{f'(x) = -\frac{2}{3\sqrt[3]{x^4}} - 3sen\, x}$$

$$y = 3x(6x - 5x^2)$$

038

The function is rewritten before derivation

$$y = 18x^2 - 15x^3$$

$$y' = 36x - 45x^2$$

$$\boxed{y' = \mathbf{36x - 45x^2}}$$

$$f(x) = \frac{x^2}{(x^2 + 5)^{-2}}$$

039

Before deriving, the function is rewritten by solving the square of the sum

$$f(x) = x^2(x^2 + 5)^2 = x^2(x^4 + 10x^2 + 25)$$

$$f(x) = x^6 + 10x^4 + 25x^2 \quad \boxed{f'(x) = \mathbf{6x^5 + 40x^3 + 50x}}$$

Note: The reader can appreciate the importance of rewriting the function.

040

$$p(t) = (2t)^4 - (2t)^2$$

The function is rewritten before derivation

$$p(t) = 16t^4 - 4t^2$$

$$p'(t) = 64t^3 - 8t$$

$$\boxed{p'(t) = 64t^3 - 8t}$$

041

A coin is dropped from the top of a building that has a height of **1362** feet (415.14 meters).

a) Determine the functions that describe the position and speed of the coin.

b) Calculate the average speed in the interval [1, 2].

c) Find the instantaneous velocities when t_1 = 1 s and t_2 = 2 s.

d) Calculate the time it takes to reach the ground

e) Determine your speed when falling on the ground

Solution:

a) Use the position function s (t) = $-16t^2 + v_o t + s_o$ for objects in free fall

If the coin is dropped $v_o = 0$.
s_o represents the height of the building 1362 feet.
Therefore, the position function is S (t) = $-16t^2 + 1362$

To determine the speed function, the position function is derived.

$$\frac{ds(t)}{dt} = -16\frac{d}{dt}(t^2) + \frac{d}{dt}(1362)$$

$$\frac{ds(t)}{dt} = v(t) = -32t$$

b) Average speed (V.P.) for the interval [1, 2].

$$\text{V. P.} = \frac{s(2)-s(1)}{2-1} = \frac{\left[-16(2)^2+1362\right]-\left[-16(1)^2+1362\right]}{2-1}$$

$$V.P. = -48\frac{feet}{s}$$

Note: The speed is positive when an object is raised and negative when it descends.

c) To make the respective calculation, the speed function is used.

$$V(t) = -32t$$

$$V(1) = -32(1) = -32 \text{ feet/s}$$

$$V(2) = -32(2) = -64 \text{ feet/s.}$$

d) To calculate the moment in which the coin touches the ground, the po function is made equal to zero and t is cleared.

$$S(t) = -16t^2+1362 = 0$$

$$-16t^2 + 1362 = 0$$

$$1362 = 16t^2 \rightarrow t = \sqrt{\frac{1362}{16}} \rightarrow t = 9.23 \text{ s}$$

e) To calculate the speed, you work with the speed function.

$$V(t) = -32t$$

For a t = 9.23 s (Time when the coin touches the ground).

$$V(9.23) = -32(9.23) = -295.36 \text{ feet/s.}$$

From a height of **220 feet**, a ball with an initial velocity of **-22 feet/s** is thrown down.
What is your speed at 3 seconds? And after descending 108 feet?

042

Also, use the position function of the previous problem. That is to say

$$S(t) = -16t^2 + v_o t + s_o$$

To answer the first question it is necessary to determine the function of position, that is

$$S(t) = -16t^2 + v_o t + s_o$$

$$S(t) = -16t^2 - 22t + 220$$

$$\frac{ds(t)}{dt} = -16\frac{d}{dt}(t^2) - 22\frac{d}{dt}(t) + \frac{d}{dt}(220)$$

$$\frac{ds(t)}{dt} = v(t) = -32t - 22$$

Then proceed to calculate the speed at 3 seconds.

$$v(3) = -32t - 22 = -32(3) - 22 = -96 - 22$$

$$\boxed{v(3) = -118\ feet/s}$$

After descending 108 feet you have

$$S(t) = -16t^2 - 22t + 108$$

$$-16t^2 - 22t + 108 = 0\ \text{(multiply by -1)}$$

$$16t^2 + 22t - 108 = 0$$

When applying the mathematical solvent one has the time of 2 seconds.

Then we proceed to calculate the speed V (2)

$$V(2) = -32t-22 = -32(2)-22 = -64-22$$

$$\boxed{V(2) = -86\ \text{feet/s}}$$

. . . .

DIFFERENTIATION RULE: Product rule

. . . .

If f and g are differentiable functions, then

$$\frac{d}{dx}[f(x)\,g(x)] = f(x)\,g'(x) + g(x)\,f'(x)$$

The rule of the product is usually stored verbally in the following way: The first function by the derivative of the second function plus the second function by the derivative of the first.

$$y = (x^3 - 2x^2 + 4)(8x^2 + 5x)$$

043

$$\frac{dy}{dx} = (x^3 - 2x^2 + 4)\frac{d}{dx}(8x^2 + 5x) + (8x^2 + 5x)\frac{d}{dx}(x^3 - 2x^2 + 4)$$

$$= (x^3 - 2x^2 + 4)(16x + 5) + (8x^2 + 5x)(3x^2 - 4x)$$

$$= 16x^4 - 32x^3 + 64x + 5x^3 - 10x^2 + 20 + 24x^4 + 15x^3 - 32x^3 - 20x^2$$

$$= 40x^4 - 44x^3 - 30x^2 + 64x + 20$$

$$\boxed{\frac{dy}{dx} = 40x^4 - 44x^3 - 30x^2 + 64x + 20}$$

$$G(x) = (x^2 + 1)(x^2 - 2x)$$

044

$$\frac{dG(x)}{dx} = (x^2 + 1)\frac{d}{dx}(x^2 - 2x) + (x^2 - 2x)\frac{d}{dx}(x^2 + 1)$$

$$= (x^2 + 1)(2x - 2) + (x^2 - 2x)(2x)$$

$$= 2x^3 + 2x - 2x^2 - 2 + 2x^3 - 4x^2$$

$$= 4x^3 - 6x^2 + 2x - 2$$

$$\boxed{\frac{dG(x)}{dx} = 4x^3 - 6x^2 + 2x - 2}$$

$$f(x) = (6x + 5)(x^3 - 2)$$

045

$$f'(x) = (6x + 5)\frac{d}{dx}(x^3 - 2) + (x^3 - 2)\frac{d}{dx}(6x + 5)$$

$$= (6x + 5)(3x^2) + (x^3 - 2)(6)$$

$$= 18x^3 + 15x^2 + 6x^3 - 12$$

$$= 24x^3 + 15x^2 - 12$$

$$\boxed{f'(x) = 24x^3 + 15x^2 - 12}$$

046 $\qquad\qquad\qquad\qquad$ $\boxed{g(s) = \sqrt{s}\,(4 - s^2)}$

$$g(s) = s^{\frac{1}{2}}(4 - s^2)$$

$$g'(s) = s^{\frac{1}{2}}\frac{d}{ds}(4 - s^2) + (4 - s^2)\frac{d}{ds}s^{\frac{1}{2}}$$

$$= s^{\frac{1}{2}}(-2s) + (4 - s^2)\frac{1}{2}\,s^{-\frac{1}{2}} \rightarrow \textit{Proceed to simplify}$$

$$= -2s\sqrt{s} + \frac{4 - s^2}{2\sqrt{s}}$$

$$= \frac{(-2s\sqrt{s})(2\sqrt{s}) + (4 - s^2)}{2\sqrt{s}}$$

$$= \frac{-4s\,s + 4 - s^2}{2\sqrt{s}}$$

$$= \frac{-4s^2 + 4 - s^2}{2\sqrt{s}}$$

$$= \frac{-5s^2 + 4}{2\sqrt{s}}$$

$$\boxed{g'(s) = \frac{4 - 5s^2}{2\sqrt{s}}}$$

$$h(t) = \sqrt[3]{t}\ (t^2 + 4)$$

047

$$h(t) = t^{\frac{1}{3}}\ (t^2 + 4)$$

$$h'(t) = t^{\frac{1}{3}}\ \frac{d}{dx}(t^2 + 4) + (t^2 + 4)\frac{d}{dx}t^{\frac{1}{3}}$$

$$= t^{\frac{1}{3}}(2t) + (t^2 + 4)\frac{1}{3}t^{-\frac{2}{3}} \quad \rightarrow \textit{Proceed to simplify}$$

$$= 2t^{\frac{4}{3}} + \frac{1}{3}\ t^{\frac{4}{3}} + \frac{4}{3}t^{-\frac{2}{3}}$$

$$= t^{\frac{4}{3}}\left(2 + \frac{1}{3}\right) + \frac{4}{3}t^{-\frac{2}{3}}$$

$$= \sqrt[3]{t^4}\ \left(\frac{7}{3}\right) + \frac{4}{3}\ t^{-\frac{2}{3}}$$

$$= \frac{7\sqrt[3]{t^4}}{3} + \frac{4}{3\sqrt[3]{t^2}}$$

$$= \frac{7\sqrt[3]{t^4}\ \sqrt[3]{t^2} + 4}{3\sqrt[3]{t^2}}$$

$$= \frac{7\sqrt[3]{t^6} + 4}{3\sqrt[3]{t^2}} \quad \leftarrow \textit{the root is simplified}$$

$$= \frac{7t^2 + 4}{3\sqrt[3]{t^2}}$$

$$\boxed{h'(t) = \frac{7t^2 + 4}{3\sqrt[3]{t^2}}}$$

048

$$f(x) = x^3 \cos x$$

$$f'(x) = x^3 \frac{d}{dx}(\cos x) + \cos x \frac{d}{dx}(x^3)$$

$$f'(x) = -x^3 sen\, x + 3x^2 \cos x \quad \leftarrow \text{ The derivative of cosx Is -senx}$$

$$= x^2 (3 \cos x - xsen\, x)$$

$$\boxed{f'(x) = x^2 (3 \cos x - xsen\, x)}$$

049

$$g(x) = \sqrt{x}\, sen\, x$$

$$g(x) = x^{\frac{1}{2}} sen\, x$$

$$g'(x) = x^{\frac{1}{2}} \frac{d}{dx}(sen\, x) + sen\, x \frac{d}{dx}\left(x^{\frac{1}{2}}\right)$$

$$g'(x) = \cos x\; x^{\frac{1}{2}} + \frac{1}{2} sen\, x\; x^{-\frac{1}{2}} \quad \leftarrow \text{ The derivative of the senx is cosx}$$

$$= \sqrt{x}\, \cos x + \frac{sen\, x}{2\sqrt{x}}$$

$$= \frac{2\sqrt{x}\; \sqrt{x}\cos x + sen\, x}{2\sqrt{x}}$$

$$= \frac{2x\cos x + sen\, x}{2\sqrt{x}}$$

$$\boxed{g'(x) = \frac{2x\cos x + sen\, x}{2\sqrt{x}}}$$

$$y = (4x + 1)(2x^2 - x)(x^3 - 8x)$$

The first two factors are identified as the first function, ie

$$(4x + 1)(2x^2 - x) \quad \leftarrow \quad \textit{the first function}$$

$$\frac{dy}{dx} = (4x + 1)(2x^2 - x)\frac{d}{dx}(x^3 - 8x) + (x^3 - 8x)\frac{d}{dx}(4x + 1)(2x^2 - x)$$

$$= (4x + 1)(2x^2 - x)(3x^2 - 8) + (x^3 - 8x)[(4x + 1)(4x - 1) + (2x^2 - x)(4)]$$

The algebraic expression that is in the brackets is the result of having applied the product rule again.

Then proceed to resolve the expressions indicated in the brackets

$$\frac{dy}{dx} = (4x + 1)(2x^2 - x)(3x^2 - 8) + (x^3 - 8x)[16x^2 - 1 + 8x^2 - 4x]$$

$$= (4x + 1)(2x^2 - x)(3x^2 - 8) + (x^3 - 8x)(24x^2 - 4x - 1)$$

$$\boxed{y' = (4x + 1)(2x^2 - x)(3x^2 - 8) + (x^3 - 8x)(24x^2 - 4x - 1}$$

The radius of a circular straight cylinder is given by $\sqrt{t + 2}$ and its height by $\frac{1}{2}\sqrt{t}$, where t is the time in seconds and the dimensions are in inches. Find the rate of volume change with respect to time $\left(\frac{dv}{dt}\right)$.

The volume of a circular straight cylinder is

$$V = \pi r^2 h$$

Where: $r = \sqrt{t + 2}$, $h = \frac{1}{2}\sqrt{t}$, $\pi = 3,1415$

Substituting r and h in the volume formula

$$V = \pi r^2 h = \pi(\sqrt{t + 2})^2 \frac{1}{2}\sqrt{t}$$

$$V = \frac{\pi}{2}(t+2)\sqrt{t}$$

The volume is derived with respect to time:

$$\frac{dv}{dt} = \frac{\pi}{2}\left[(t+2)\frac{dv}{dt}\left(t^{\frac{1}{2}}\right) + t^{\frac{1}{2}}(\frac{dv}{dt}(t+2))\right]$$

$$= \frac{\pi}{2}\left[\frac{1}{2}(t+2)t^{-\frac{1}{2}} + t^{\frac{1}{2}}(1)\right]$$

$$= \frac{\pi}{2}\left[\frac{t+2}{2\sqrt{t}} + \sqrt{t}\right]$$

$$= \frac{\pi}{2}\left[\frac{t+2+2t}{2\sqrt{t}}\right]$$

$$= \frac{\pi}{2}\left[\frac{3t+2}{2\sqrt{t}}\right]$$

$$\boxed{\frac{dv}{dt} = \pi\left(\frac{3t+2}{4\sqrt{t}}\right) \; inch^3/s}$$

052

The length of a rectangle is given by **2t+1** and its height is $\sqrt{t}$, where **t** is the time in seconds and the dimensions are in centimeters. **Find the rate of change of the area with respect to time** $\frac{dA}{dt}$.

The area of the rectangle is represented by the following formula

$$A = L\,h$$

Where L is the length y is equal to 2t + 1. And h the height and is equal to $\sqrt{t}$

Substituting L and h in the formula of the area, you have

$$A = (\,2t+1)\,\sqrt{t}$$

Then proceed to derive the area with respect to time

$$\frac{dA}{dt} = \frac{d}{dt} \left[(2t+1) \right] t^{\frac{1}{2}}$$

$$= (2t+1) \left(\frac{1}{2} t^{-\frac{1}{2}} \right) + 2\, t^{\frac{1}{2}}$$

$$= \frac{2t+1}{2t^{\frac{1}{2}}} + 2\, t^{\frac{1}{2}}$$

$$= \frac{2t+1+(2t^{\frac{1}{2}})(2t^{\frac{1}{2}})}{2t^{\frac{1}{2}}}$$

$$= \frac{2t+1+4t}{2t^{\frac{1}{2}}}$$

$$= \frac{6t+1}{2t^{\frac{1}{2}}}$$

$$\boxed{\frac{dA}{dt} = \frac{6t+1}{2\sqrt{t}}\ cm^2/s}$$

. . . .

DIFFERENTIATION RULE: Quotient rule

. . . .

If f and g are differentiable functions and g (x) ≠ 0, then

$$\frac{d}{dx} \left[\frac{f(x)}{g(x)} \right] = \frac{g(x)f'(x) - f(x)g'(x)}{[g(x)]^2}$$

In verbal form, the quotient rule is: The denominator by the derivative of the numerator, minus the numerator multiplied by the derivative of the denominator, all divided by the denominator squared.

In the following exercises, you will find the derivative of each one of the functions.

053

$$f(x) = \frac{x}{x^2 + 1}$$

For the resolution of the problem posed, the use of the formula described above is required. That is to say

$$f'(x) = \frac{(x^2+1)\frac{d}{dx}(x) - x\frac{d}{dx}(x^2+1)}{(x^2+1)^2}$$

$$= \frac{(x^2+1)(1) - x(2x)}{(x^2+1)^2} \quad \leftarrow \textit{It proceeds to simplify}$$

$$= \frac{x^2+1-2x^2}{(x^2+1)^2}$$

$$= \frac{-x^2+1}{(x^2+1)^2}$$

$$= \frac{1-x^2}{(x^2+1)^2} \quad \boxed{f'(x) = \frac{1-x^2}{(x^2+1)^2}}$$

054

$$h(x) = \frac{\sqrt[3]{x}}{x^3 + 1}$$

$$h'(x) = \frac{(x^3+1)\frac{d}{dx}\left(x^{\frac{1}{3}}\right) - x^{\frac{1}{3}}\frac{d}{dx}(x^3+1)}{(x^3+1)^2} \quad \leftarrow \textit{the derivative of a quotient is applied}$$

$$h'(x) = \frac{(x^3+1)\left(\frac{1}{3}x^{-\frac{2}{3}}\right) - x^{\frac{1}{3}}(3x^2)}{(x^3+1)^2} \quad \leftarrow \textit{it proceeds to simplify}$$

66

$$h'(x) = \frac{\frac{(x^3+1)}{3\sqrt[3]{x^2}} - (3x^2)\sqrt[3]{x}}{(x^3+1)^2}$$

$$= \frac{\frac{(x^3+1) - \left(3x^2\sqrt[3]{x}\right)\left(3\sqrt[3]{x^2}\right)}{3\sqrt[3]{x^2}}}{(x^3+1)^2}$$

$$= \frac{\frac{(x^3+1) - 9x^2\sqrt[3]{x^3}}{3\sqrt[3]{x^2}}}{(x^3+1)^2} = \frac{\frac{x^3+1-9x^3}{3\sqrt[3]{x^2}}}{(x^3+1)^2}$$

$$= \frac{\frac{1-8x^3}{3\sqrt[3]{x^2}}}{(x^3+1)^2} = \frac{1-8x^3}{\left(3\sqrt[3]{x^2}\right)\left((x^3+1)^2\right)}$$

$$\boxed{h'(x) = \frac{1 - 8x^3}{\left(3\sqrt[3]{x^2}\right)(x^3 + 1)^2}}$$

$$\boxed{g(t) = \frac{t^2 + 2}{2t - 7}}$$

$$g'(t) = \frac{(2t-7)\frac{d}{dt}(t^2+2) - (t^2+2)\frac{d}{dt}(2t-7)}{(2t-7)^2}$$

$$= \frac{(2t-7)(2t) - (t^2+2)(2)}{(2t-7)^2} \quad \leftarrow \textit{It proceeds to simplify}$$

$$= \frac{4t^2-14t-2t^2-4}{(2t-7)^2}$$

$$= \frac{2t^2 - 14t - 4}{(2t-7)^2}$$

$$\boxed{g'(t) = \frac{2t^2 - 14t - 4}{(2t - 7)^2}}$$

056

$$\boxed{g(x) = \frac{\text{sen } x}{x^2}}$$

$$g'(x) = \frac{x^2 \frac{d}{dx}(\text{sen } x) - \text{sen } x \frac{d}{dx}(x^2)}{(x^2)^2}$$

$$= \frac{x^2 \cos x - 2x \text{sen } x}{x^4} \quad \leftarrow \textit{It proceeds to simplify}$$

$$= \frac{x(x\cos x - 2\text{sen } x)}{x^4} \quad \leftarrow \textit{Common factor x is taken out}$$

$$= \frac{x\cos x - 2\text{sen } x}{x^3}$$

$$\boxed{g'(x) = \frac{x\cos x - 2\text{sen } x}{x^3}}$$

057

$$\boxed{f(t) = \frac{\cos t}{t^3}}$$

$$f'(x) = \frac{t^3 \frac{d}{dt}(\cos t) - \cos t \frac{d}{dt}(t^3)}{(t^3)^2}$$

$$= \frac{-t^3 \operatorname{sen} t - 3\cos t (t^2)}{t^6} \quad \leftarrow \textit{it proceeds to simplify}$$

$$= \frac{-t^3 \operatorname{sen} t - 3t^2 \cos t}{t^6}$$

$$= \frac{-t^2 (t \operatorname{sen} t + 3\cos t)}{t^6} \quad \leftarrow \textit{Common factor } -t^2 \textit{ is taken out}$$

$$\boxed{f'(t) = -\frac{t \operatorname{sen} t + 3\cos t}{t^4}}$$

Boyle's law states that, if the temperature of a gas remains constant, its pressure is inversely proportional to its volume. Use the derivative to show that the rate of change of the pressure is inversely proportional to the square of the volume.

058

The law establishes that the pressure is inversely proportional to the volume, that is to say

$$P \propto \frac{1}{V}$$

To transform the proportionality into an equality, a constant (K) is introduced.

$$p = \frac{K}{V} \quad \rightarrow \quad p = kV^{-1} \quad \rightarrow \quad \frac{dp}{dv} = -kV^{-2}$$

$$\rightarrow \quad \frac{dp}{dv} = -\frac{k}{V^2}$$

059

A population of 500 bacteria is introduced into a culture and increases in number according to the equation:

$$P(t) = 500\left(1 + \frac{4t}{50+t^2}\right)$$

Where **t** is measured in hours. **Calculate the rate of change to which the population is growing when t equals 2 hours.**

P is derived with respect to time

$$\frac{dp}{dt} = \frac{d}{dt}500\left(1 + \frac{4t}{50+t^2}\right)$$

$$= 500\left[0 + \frac{(50+t^2)(4)-(4t)(2t)}{(50+t^2)^2}\right]$$

$$= 500\left[\frac{200+4t^2-8t^2}{(50+t^2)^2}\right]$$

$$= 500\left[\frac{-4t^2+200}{(50+t^2)^2}\right]$$

For a time of 2 hours you get

$$\frac{dp}{dt} = 500\left(\frac{-16+200}{2916}\right)$$

$$\frac{dp}{dt} = 500\left(\frac{184}{2916}\right)$$

$$\boxed{\frac{dp}{dt} = 31.55\ \boldsymbol{bacteria/h}}$$

DERIVATION OF TRIGONOMETRIC FUNCTIONS

Known the derivatives of the sine and cosine functions (Demonstration made in problems 8 and 9), the quotient rule allows to establish those of the four remaining trigonometric functions.

Prove that the derivative of the tangent of **x** is the secant squared of **x**.

Whereas the $tan\ x = \dfrac{sen\ x}{cos\ x}$ and applying the quotient rule you get

$$\frac{d}{dx}\left[tan\ x\right] = \frac{d}{dx}\left[\frac{sen\ x}{cos\ x}\right]$$

$$= \frac{cos\ x \dfrac{d}{dx}(sen\ x) - sen\ x \dfrac{d}{dx}(cos\ x)}{(cos\ x)^2}$$

$$= \frac{cos\ x\ cos\ x - sen\ x\ (-sen\ x)}{(cos\ x)^2}$$

$$= \frac{cos^2 x + sen^2 x}{cos^2 x}$$

As the $sen^2 x + cos^2 x = 1$

$$\frac{d}{dx}\left[tan\ x\right] = \frac{1}{cos^2 x}$$

Como $\dfrac{1}{cos\ x} = sec x$

$$\frac{d}{dx}\left[tan\ x\right] = sec^2 x$$

$$\boxed{\frac{d}{dx}\left[\boldsymbol{tan\ x}\right] = \boldsymbol{sec^2 x}}$$

061

Show that the derivative of the cotangent of x is minus the square cosecant of x.

Whereas the $\cot x = \dfrac{\cos x}{\operatorname{sen} x}$ and applying the quotient rule you get

$$\frac{d}{dx}\left[\cot x\right] = \frac{d}{dx}\left[\frac{\cos x}{\operatorname{sen} x}\right]$$

$$= \frac{\operatorname{sen} x \frac{d}{dx}(\cos x) - \cos x \frac{d}{dx}(\operatorname{sen} x)}{(\operatorname{sen} x)^2}$$

$$= \frac{\operatorname{sen} x \,(-\operatorname{sen} x) - \cos x \cos x}{(\operatorname{sen} x)^2}$$

$$= -\frac{\operatorname{sen}^2 x + \cos^2 x}{\operatorname{sen}^2 x} \qquad = -\frac{1}{\operatorname{sen}^2 x}$$

As $\dfrac{1}{\operatorname{sen} x} = \csc x$

$$\frac{d}{dx}\left[\cot x\right] = -\csc^2 x \qquad \boxed{\frac{d}{dx}\left[\cot x\right] = -csc^{\,2}x}$$

062

Show that the derivative of the secant of x is the secant of x by the tangent of x.

Whereas the $\sec x = \dfrac{1}{\cos x}$ and applying the quotient rule you get

$$\frac{d}{dx}\left[\sec x\right] = \frac{d}{dx}\left[\frac{1}{\cos x}\right]$$

$$= \frac{\cos x \frac{d}{dx}(1) - (1)\frac{d}{dx}(\cos x)}{(\cos x)^2}$$

$$= \frac{\cos x \,(0) - (1)\,(-\operatorname{sen} x)}{(\cos x)^2}$$

$$= \frac{sen\, x}{(\cos x)^2}$$

$$= \frac{sen\, x}{\cos x} \quad \frac{1}{\cos x}$$

$$= tan\, x \ \sec x \qquad \boxed{\frac{d}{dx}\left[\sec x\right] = \sec x \tan x}$$

Show that the derivative of the cosecant of **x** is minus the cosecant of **x** by the cotangent of **x**.

Whereas the $\csc x = \dfrac{1}{sen\, x}$ y and applying the quotient rule you get

$$\frac{d}{dx}\left[\csc x\right] = \frac{d}{dx}\left[\frac{1}{sen\, x}\right]$$

$$= \frac{sen\, x \, \dfrac{d}{dx}(1) - (1)\, \dfrac{d}{dx}(sen\, x)}{(sen\, x)^2}$$

$$= \frac{sen\, x\,(0) - (1)\cos x}{(sen\, x)^2}$$

$$= \frac{-\cos x}{(sen\, x)^2}$$

$$= \frac{-\cos x}{sen\, x} \quad \frac{1}{sen\, x}$$

$$= -\csc x \cot x$$

$$\boxed{\frac{d}{dx}\left[\csc x\right] = -\csc x \cot x}$$

064

$$f(t) = t^2 sen\, t$$

$$f'(t) = t^2 \frac{d}{dt}(sen\, t) + sen\, t \frac{d}{dt}(t^2) \leftarrow derived\ from\ a\ product$$

$$= t^2 \cos t + sent\, 2t$$

$$= t^2 \cos t + 2t\, sent$$

$$= t(t\, cost + 2\, sent)$$

$$\boxed{f'(t) = t(t\, cost + 2\, sent)}$$

065

$$f(\theta) = (\theta + 1)\cos\theta$$

$$f(\theta) = \theta\cos\theta + \cos\theta$$

$$f'(\theta) = \theta\frac{d}{d\theta}\cos\theta + \cos\theta\frac{d}{d\theta}(\theta) + \frac{d}{d\theta}(\cos\theta)$$

$$= -\theta\, sen\,\theta + \cos\theta(1) - sen\,\theta$$

$$= -\theta\, sen\,\theta - sen\,\theta + \cos\theta \leftarrow Common\ factor\text{-}sen\ \theta$$

$$= -sen\,\theta(\theta + 1) + \cos\theta$$

$$\boxed{f'(\theta) = -(\theta + 1)sen\,\theta + \cos\theta}$$

066

$$f(t) = \frac{\cos t}{t}$$

$$f'(x) = \frac{t\frac{d}{dt}(\cos t) - \cos t\frac{d}{dt}(t)}{t^2} \leftarrow derived\ from\ a\ quotient$$

$$f'(x) = \frac{-t\, sen\, t - \cos t}{t^2}$$

$$f'(x) = -\frac{t\, sen\, t + \cos t}{t^2} \quad \leftarrow \textit{Common factor less (-)}$$

$$\boxed{\boldsymbol{f'(x)} = -\frac{\boldsymbol{t\, sen\, t + \cos t}}{\boldsymbol{t^2}}}$$

$$\boxed{f(x) = \frac{sen\, x}{x}}$$ 067

$$f'(x) = \frac{x\frac{d}{dx}(sen\, x) - sen\, x\frac{d}{dx}(x)}{x^2}$$

$$= \frac{x\cos x - sen\, x\,(1)}{x^2}$$

$$= \frac{x\cos x - sen\, x}{x^2} \qquad \boxed{f'(x) = \frac{x\cos x - sen\, x}{x^2}}$$

$$\boxed{f(x) = -x + \tan x}$$ 068

$$f'(x) = \frac{d}{dx}(-x) + \frac{d}{dx}(\tan x)$$

$$= -1 + \sec^2 x$$

$$= \sec^2 x - 1$$

As $1 + \tan^2 x = \sec^2 x$

$$f'(x) = \tan^2 x \qquad \boxed{\boldsymbol{f'(x) = \tan^2 x}}$$

77

069

$$Y = x + \cot x$$

$$Y' = \frac{d}{dx}(x) + \frac{d}{dx}(\text{Cot } x)$$

$$= 1 + (-\csc^2 x)$$

$$= 1 - \csc^2 x$$

As $1 + \cot^2 x = \csc^2 x$

$$Y' = -\cot^2 x \qquad \boxed{\boldsymbol{Y' = -\cot^2 x}}$$

070

$$g(t) = \sqrt[4]{t} + 8\sec t$$

$$g(t) = t^{\frac{1}{4}} + 8\sec t \quad \leftarrow \textit{Function is rewritten}$$

$$g'(t) = \frac{d}{dt}\left(t^{\frac{1}{4}}\right) + 8\frac{d}{dt}(\sec t)$$

$$= \frac{1}{4}t^{-\frac{3}{4}} + 8\sec t\,\tan t \leftarrow \frac{d}{dt}(\sec t) = \sec t \tan t$$

$$= \frac{1}{4t^{\frac{3}{4}}} + 8\sec t\,\tan t \qquad \boxed{\boldsymbol{g'(t) = \frac{1}{4\sqrt[4]{t^3}} + 8\sec t\,\tan t}}$$

071

$$h(s) = \frac{1}{s} - 10\csc s$$

$$h'(s) = \frac{d}{ds}(s^{-1}) - 10\frac{d}{ds}(\csc s) \leftarrow \textit{Function is rewritten}$$

$$= -1s^{-2} - 10(-\csc s \cot s) \leftarrow \frac{d}{ds}(\csc s) = -\csc s \cot s$$

$$= -\frac{1}{s^2} + 10 \ \csc s \cot s$$

$$\boxed{h'(s) = -\frac{1}{s^2} + 10 \ \csc s \cot s}$$

$$\boxed{Y = \frac{3(1 - sen \ x)}{2 \cos x}} \qquad \fbox{072}$$

$$Y = \frac{3 - 3sen \ x}{2 \cos x} \quad \leftarrow \ \textit{Function is rewritten}$$

$$Y' = \frac{2\cos x \frac{d}{dx}(3 - 3sen \ x) - (3 - 3sen \ x)\frac{d}{dx}(2\cos x)}{(2\cos x)^2}$$

$$= \frac{2\cos x(-3\cos x) - [(3 - 3sen \ x)(-2\sen x)]}{4cos^2 x}$$

$$= \frac{-6cos^2 x - [-6sen \ x + 6sen^2 x]}{4cos^2 x}$$

$$= \frac{-6cos^2 x + 6sen \ x - 6sen^2 x}{4cos^2 x}$$

$$= \frac{6(-cos^2 x + sen \ x - sen^2 x)}{4cos^2 x}$$

$$= \frac{3}{2}\left[-\frac{cos^2 x}{cos^2 x} + \frac{sen \ x}{\cos x}\frac{1}{\cos x} - \frac{sen^2 x}{cos^2 x}\right]$$

$$= \frac{3}{2}\left[-1 + \sec x \tan x - tan^2 x\right]$$

$$= \frac{3}{2}\left[-(1 + tan^2 x] + \sec x \tan x\right]$$

$$= \frac{3}{2}(-sec^2 x + \sec x \tan x) = \frac{3}{2}\sec x(\tan x - \sec x)$$

$$\boxed{Y' = \frac{3}{2}\sec x(\tan x - \sec x)}$$

073

$$Y = \frac{\sec x}{x}$$

$$Y' = \frac{x\frac{d}{dx}(\sec x) - \sec x\frac{d}{dx}(x)}{x^2} = \frac{x(\sec x \tan x) - \sec x}{x^2} = \frac{x \sec x \tan x - \sec x}{x^2}$$

$$= \frac{\sec x(x \tan x - 1)}{x^2}$$

$$\boxed{Y' = \frac{\sec x(x \tan x - 1)}{x^2}}$$

074

$$Y = -\csc x - sen\, x$$

$$y' = -\frac{d}{dx}(\csc x) - \frac{d}{dx}(sen\, x)$$

$$= -(-\csc x \cot x) - \cos x$$

$$= \csc x \cot x - \cos x$$

$$= \frac{1}{sen\, x}\frac{\cos x}{sen\, x} - \cos x$$

$$= \frac{\cos x}{sen^2 x} - \cos x$$

$$= \cos x\left(\frac{1}{sen^2 x} - 1\right)$$

$$= \cos x(csc^2 x - 1)$$

$$= \cos x\, cot^2 x \qquad \boxed{y' = \cos x\, cot^2 x}$$

$$Y = x\,sen\,x + \cos x$$

075

The first term of the function is a product; the derivative of a product is applied

$$Y' = x\frac{d}{dx}(sen\,x) + sen\,x\frac{d}{dx}(x) + \frac{d}{dx}(\cos x)$$

$$= x\cos x + sen\,x\,(1) - sen\,x$$

$$= x\cos x + sen\,x - sen\,x \quad \boxed{Y' = x\cos x}$$

$$f(x) = x^2 \tan x$$

076

The derivative of a product is applied

$$f'(x) = x^2\frac{d}{dx}(\tan x) + \tan x\frac{d}{dx}(x^2)$$

$$= x^2 sec^2 x + \tan x(2x)$$

$$= x^2 sec^2 x + 2x\tan x$$

$$= x(x\,sec^2 x + 2\tan x)$$

$$f'(x) = x(x\,sec^2 x + 2\tan x) \quad \boxed{f'(x) = x(x\,sec^2 x + 2\tan x)}$$

$$f(x) = sen\,x\cos x$$

077

$$f'(x) = sen\,x\frac{d}{dx}(\cos x) + \cos x\frac{d}{dx}(sen\,x)$$

$$= sen\,x\,(-sen\,x) + \cos x\cos x$$

$$= -sen^2 x + cos^2 x$$

81

$$= cos^2x - sen^2x$$

As $cos\, 2x = cos^2x - sen^2x$

$$f'(x) = cos\, 2x \qquad \boxed{f'(x) = \textbf{cos\, 2x}}$$

078

$$\boxed{Y = 2x\, sen\, x + x^2 \cos x}$$

$$Y' = 2x\frac{d}{dx}(sen\, x) + sen\, x\frac{d}{dx}(2x) + x^2\frac{d}{dx}(\cos x) + \cos x\frac{d}{dx}(x^2)$$

$$= 2x\cos x + sen\, x(2) + x^2(-sen\, x) + \cos x(2x)$$

$$= 2x\cos x + 2sen\, x - x^2 sen\, x + 2x\cos x$$

$$= 4x\cos x + sen\, x(2 - x^2)$$

$$Y' = 4x\cos x + sen\, x(2 - x^2)$$

$$\boxed{Y' = \textbf{4x cos x} + \textbf{sen x}(\textbf{2} - \textbf{x}^2)}$$

079

$$\boxed{h(\theta) = 5\theta \sec\theta + \theta \tan\theta}$$

The derivative of a product is applied to each term of the function

$$h'(\theta) = 5\left[\theta\frac{d}{d\theta}(sec\theta) + \sec\theta\frac{d}{d\theta}(\theta)\right] + \theta\frac{d}{d\theta}(\tan\theta) + \tan\theta\frac{d}{d\theta}(\theta)$$

$$= 5\theta \sec\theta \tan\theta + 5\sec\theta(1) + \theta sec^2\theta + tan\theta(1)$$

$$h'(\theta) = 5\theta \sec\theta \tan\theta + 5\sec\theta + \theta sec^2\theta + tan\theta$$

$$\boxed{h'(\theta) = \textbf{5}\theta \sec\theta \tan\theta + \textbf{5}\sec\theta + \theta sec^2\theta}$$

$$Y = (x^2 + sen\ x)\sec\ x$$

080

The function is rewritten by solving the indicated product

$$Y = x^2 \sec x + sen\ x\sec x$$

$$Y' = x^2 \frac{d}{dx}\left(\sec x\right) + \sec x\frac{d}{dx}\left(x^2\right)\right) + sen\ x\frac{d}{dx}\left(\sec x\right) + \sec x\frac{d}{dx}\left(sen\ x\right)$$

$$= x^2 \sec x\tan x + 2x\sec x + sen\ x\ \sec x\tan x + \sec x\cos x$$

$$= x^2 \sec x\tan x + 2x\sec x + sen\ x\ \frac{1}{\cos x}\tan x + \frac{1}{\cos x}\cos x$$

$$= x^2 \sec x\tan x + 2x\sec x + tan^2 x + 1$$

As $1 + tan^2 x = sec^2 x$

$$Y' = x^2 \sec x\tan x + 2x\sec x + sec^2 x$$

$$\boxed{Y' = x^2 \sec x\tan x + 2x\sec x + sec^2 x}$$

$$f(x) = \frac{\cot x}{x+1}$$

081

$$f'(x) = \frac{(x+1)\frac{d}{dx}(\cot x) - \cot x\frac{d}{dx}(x+1)}{(x+1)^2}$$

$$f'(x) = \frac{(x+1)(-csc^2 x) - \cot x\,(1)}{(x+1)^2} \quad \leftarrow \frac{d}{dx}[\cot x] = -csc^2 x$$

$$f'(x) = \frac{-x\,csc^2 x - csc^2 x - \cot x}{(x+1)^2}$$

$$f'(x) = -\frac{x\,csc^2 x + csc^2 x + \cot x}{(x+1)^2} \qquad \boxed{f'(x) = -\frac{csc^2 x(x+1) + \cot x}{(x+1)^2}}$$

082

$$Y = \frac{x^2}{1 + 2\tan x}$$

$$Y' = \frac{(1+2\tan x)\frac{d}{dx}\left(x^2\right) - x^2 \frac{d}{dx}(1+2\tan x)}{(1+2\tan x)^2}$$

$$= \frac{(1+2\tan x)(2x) - x^2(2\sec^2 x)}{(1+2\tan x)^2} \quad \leftarrow \frac{d}{dx}[\tan x] = \sec^2 x$$

$$= \frac{4x\tan x - 2x^2\sec^2 x + 2x}{(1+2\tan x)^2} \qquad \boxed{Y' = \frac{4x\tan x - 2x^2\sec^2 x + 2x}{(1 + 2\tan x)^2}}$$

083

$$Y = x^3 \cos x - x^3 sen\, x$$

The derivative of a product is applied to both terms of the function

$$Y' = x^3\frac{d}{dx}(\cos x) + \cos x\frac{d}{dx}(x^3) - \left[x^3\frac{d}{dx}(sen\, x) + sen\, x\frac{d}{dx}(x^3)\right]$$

$$= x^3(-sen\, x) + \cos x(3x^2) - [x^3(\cos x) + sen\, x(3x^2)]$$

$$= -x^3\, sen\, x + 3x^2\cos x - x^3\cos x - 3x^2 sen\, x$$

$$= 3x^2\cos x - x^3\, sen\, x - x^3\cos x - 3x^2 sen\, x$$

$$= x^2(3\cos x - x\, sen\, x) - x^2(x\cos x + 3\, sen\, x)$$

$$= x^2(3\cos x - x\, sen\, x - x\cos x - 3\, sen\, x)$$

$$= x^2(3\cos x - x\cos x - 3\, sen\, x - x\, sen\, x)$$

$$= x^2[\cos x(3 - x) - sen\, x(3 + x)]$$

$$\boxed{Y' = x^2[\cos x(3 - x) - sen\, x(3 + x)]}$$

$$Y = x^2 sen\, x \tan x$$

084

The problem presents the product of three factors. To apply the product rule we proceed to identify the first factors as the first function

$x^2 sen\, x \rightarrow$ the first function

$$Y' = (x^2 sen\, x)\frac{d}{dx}(\tan x) + \tan x \frac{d}{dx}(x^2 senx)$$

$$= (x^2 sen\, x)(sec^2 x) + \tan x[(x^2 \cos x + sen\,(2x)]$$

$$= (x^2 sen\, x)(sec^2 x) + \tan x[x^2 \cos x + 2x senx]$$

$$= x^2 sen\, x\, sec^2 x + x^2 \cos x \tan x + 2x sen\, x \tan x$$

$$= x^2 sen\, x sec^2 x + x^2 \cos x \frac{sen\, x}{\cos x} + 2x sen\, x \tan x$$

$$= x^2 sen\, x\, sec^2 x + x^2 sen\, x + 2x\, sen\, x \tan x$$

$$\boxed{Y' = x^2 sen\, x\, sec^2 x + x^2 sen\, x + 2x sen\, x \tan x}$$

Note: Note that the expression that is written in square brackets (second step) is the result of applying the product rule again.

DIFFERENTIATION RULE: Rule of the Chain

If Y = f (u) is a derivable function of u and besides u = g (x) is a derivable function of x, then Y = f (g (x)) is a derivable function of x.

$$\frac{dy}{dx} = \frac{dy}{du}\frac{du}{dx}$$

or its equivalent

$$\frac{d}{dx}[f(g(x))] = f'(g(x))\, g'(x)$$

085

$$Y = (2x - 7)^3$$

Applying the rule of the chain you have

$$Y' = \frac{d}{dx}[(2x - 7)^3]$$

$$= 3[(2x - 7)^2]\frac{d}{dx}(2x - 7)$$

$$= 3[(2x - 7)^2](2)$$

$$= 6[(2x - 7)^2] \quad \boxed{\boldsymbol{Y' = 6[(2x - 7)^2]}}$$

Observation: When the chain rule is applied, it is derived from the outside inwards.

In our case, it is derived $(2x - 7)^3$ *and then it is derived* $2x - 7$

086

$$Y = 3(4 - x^2)^5$$

$$Y' = \frac{d}{dx}[3(4 - x^2)^5]$$

$$= 15(4 - x^2)^4\frac{d}{dx}(4 - x^2)$$

$$= 15(4 - x^2)^4(-2x)$$

$$= -30x(4 - x^2)^4$$

$$\boxed{\boldsymbol{Y' = -30x(4 - x^2)^4}}$$

$$g(x) = 3(4 - 9x)^4$$

087

$$g'(x) = \frac{d}{dx}[3(4 - 9x)^4]$$

$$= 12(4 - 9x)^3 \frac{d}{dx}(4 - 9x)$$

$$= 12(4 - 9x)^3(-9)$$

$$= -108\ (4 - 9x)^3$$

$$\boxed{g'(x) = -108\ (4 - 9x)^3}$$

$$Y = \frac{1}{(x^3 - 2x^2 + 7)^4}$$

088

The function is rewritten as follows

$$Y = (x^3 - 2x^2 + 7)^{-4}$$

$$Y' = -4(x^3 - 2x^2 + 7)^{-5} \frac{d}{dx}(x^3 - 2x^2 + 7)$$

$$Y' = -4(x^3 - 2x^2 + 7)^{-5}\ (3x^2 - 4x)$$

$$\boxed{Y' = -4(x^3 - 2x^2 + 7)^{-5}\ (3x^2 - 4x)}$$

$$Y = (3x - 1)^4\ (-2x + 9)^5$$

089

$$Y' = (3x - 1)^4 \frac{d}{dx}(-2x + 9)^5 + (-2x + 9)^5 \frac{d}{dx}((3x - 1)^4$$

$$= (3x - 1)^4\ 5(-2x + 9)^4 \frac{d}{dx}(-2x + 9) + (-2x + 9)^5\ 4((3x - 1)^3)\frac{d}{dx}(3x - 1)$$

$$= (3x - 1)^4 \, 5(-2x + 9)^4(-2) + (-2x + 9)^5 \, 4 \, (3x - 1)^3(3)$$

$$= (3x - 1)^4 \, (-10)(-2x + 9)^4 + (-2x + 9)^5 \, (12) \, (3x - 1)^3$$

$$= -10(3x - 1)^4(-2x + 9)^4 + 12(-2x + 9)^5(3x - 1)^3$$

$$\boxed{Y' = 12(-2x + 9)^5(3x - 1)^3 - 10(3x - 1)^4(-2x + 9)^4}$$

090

$$f(x) = \left(\frac{x^2 - 1}{x^2 + 1}\right)^2$$

$$f'(x) = \frac{d}{dx}\left(\frac{x^2-1}{x^2+1}\right)^2$$

$$= 2\left(\frac{x^2-1}{x^2+1}\right)\frac{d}{dx}\left(\frac{x^2-1}{x^2+1}\right)$$

$$= 2\left(\frac{x^2-1}{x^2+1}\right)\left(\frac{(x^2+1)(2x)-(x^2-1)(2x)}{(x^2+1)^2}\right)$$

$$= 2\left(\frac{x^2-1}{x^2+1}\right)\left(\frac{2x^3+2x-2x^3+2x}{(x^2+1)^2}\right)$$

$$= 2\left(\frac{x^2-1}{x^2+1}\right)\left(\frac{4x}{(x^2+1)^2}\right)$$

$$= \frac{8x(x^2-1)}{(x^2+1)^3} \quad \rightarrow \quad \boxed{f'(x) = \frac{8x(x^2 - 1)}{(x^2 + 1)^3}}$$

$$H(u) = (2 + u \, sen \, u)^{-3}$$

091

$$H'(u) = \frac{d}{du}(2 + u \, sen \, u)^{-3}$$

$$= -3(2 + u \, sen \, u)^{-4}\frac{d}{dx}(2 + u \, sen \, u)$$

$$= -3(2 + u \, sen \, u)^{-4}(u \, cos \, u + sen \, u)$$

$$= \frac{-3(ucos \, u + sen \, u)}{(2+u \, sen \, u)^4} \rightarrow$$

$$\boxed{H'(u) = -\frac{3(ucos \, u + sen \, u)}{(2 + u \, sen \, u)^4}}$$

$$F(\theta) = (2\theta + 1)^3 tan^2\theta$$

092

$$F'(\theta) = (2\theta + 1)^3\frac{d}{d\theta}(\tan\theta)^2 + (\tan\theta)^2\frac{d}{dx}(2\theta + 1)^3$$

$$= (2\theta + 1)^3 2\tan\theta\frac{d}{d\theta}\tan\theta + (\tan\theta)^2 3(2\theta + 1)^2\frac{d}{d\theta}(2\theta + 1)$$

$$= (2\theta + 1)^3 2\tan\theta sec^2\theta + (\tan\theta)^2 3(2\theta + 1)^2(2)$$

$$= 2(2\theta + 1)^3\tan\theta sec^2\theta + (\tan\theta)^2 6(2\theta + 1)^2$$

$$= 2(2\theta + 1)^2\tan\theta[(2\theta + 1)sec^2\theta + 3\tan\theta]$$

$$\boxed{F'(\theta) = 2(2\theta + 1)^2\tan\theta[(2\theta + 1)sec^2\theta + 3\tan\theta]}$$

093

$$f(x) = [x + (x^2 - 4)^3]^{10}$$

$$f'(x) = 10[x + (x^2 - 4)^3]^9 \frac{d}{dx}(x + (x^2 - 4)^3)$$

$$= 10[x + (x^2 - 4)^3]^9 (1 + 3(x^2 - 4)^2 \frac{d}{dx}(x^2 - 4))$$

$$= 10[x + (x^2 - 4)^3]^9 (1 + 3(x^2 - 4)^2 (2x))$$

$$= 10[x + (x^2 - 4)^3]^9 (1 + 6x(x^2 - 4)^2)$$

$$\boxed{f'(x) = 10[x + (x^2 - 4)^3]^9 (1 + 6x(x^2 - 4)^2)}$$

094

$$y = sen\, 2x \cos 3x$$

$$Y' = sen\, 2x \frac{d}{dx}(cos\, 3x) + cos\, 3x \frac{d}{dx}(sen\, 2x)$$

$$= sen\, 2x\, (-sen3x)(3) + cos\, 3x(cos\, 2x)(2)$$

$$= -3sen\, 2x\, (sen3x) + 2cos\, 3x \cos 2x$$

$$\boxed{Y' = -3sen\, 2x\, sen\, 3x + 2cos\, 3x \cos 2x}$$

095

$$f(x) = (\sec 4x + \tan 2x)^5$$

$$f'(x) = \frac{d}{dx}(\sec 4x + \tan 2x)^5$$

$$= 5\,(\sec 4x + \tan 2x)^4 \frac{d}{dx}(\sec 4x + \tan 2x)$$

$$= 5\,(\sec 4x + \tan 2x)^4\,(4\sec 4x\, tan\, 4x + 2sec^2 2x)$$

$$\boxed{f'(x) = 5\,(\sec 4x + \tan 2x)^4\,(4\sec 4x\, tan4x + 2sec^2 2x)}$$

$$f(x) = sen(sen\ x)$$

096

$$f'(x) = \cos(sen\ x)\frac{d}{dx}(sen\ x)$$

$$= \cos(sen\ x)(\cos x)$$

$$\boxed{f'(x) = \cos(sen\ x)(\cos x)}$$

$$f(x) = (9t + 2)^{\frac{2}{3}}$$

097

$$f'(x) = \frac{d}{dx}(9t + 2)^{\frac{2}{3}}$$

$$= \frac{2}{3}(9t + 2)^{-\frac{1}{3}}\frac{d}{dx}(9t + 2)$$

$$= \frac{2}{3}(9t + 2)^{-\frac{1}{3}}(9) \qquad \boxed{f'(x) = \frac{6}{\sqrt[3]{9t + 2}}}$$

$$f(t) = \sqrt{1 - t}$$

098

$$f(t) = (1 - t)^{\frac{1}{2}} \leftarrow Function\ is\ rewritten$$

$$f'(t) = \frac{d}{dt}(1 - t)^{\frac{1}{2}}$$

$$= \frac{1}{2}(1 - t)^{-\frac{1}{2}}\frac{d}{dt}(1 - t)$$

$$= \frac{1}{2}(1-t)^{-\frac{1}{2}}(-1)$$

$$\boxed{f'(t) = -\frac{1}{2\sqrt{1-t}}}$$

099

$$\boxed{g(x) = \sqrt{5-3x}}$$

$$g(x) = (5-3x)^{\frac{1}{2}} \leftarrow \textit{Function is rewritten}$$

$$g'(x) = \frac{1}{2}(5-3x)^{-\frac{1}{2}}\frac{d}{dx}(5-3x)$$

$$= \frac{1}{2}(5-3x)^{-\frac{1}{2}}(-3)$$

$$= -\frac{3}{2\sqrt{5-3x}} \qquad \boxed{g'(x) = -\frac{3}{2\sqrt{5-3x}}}$$

100

$$\boxed{Y = \sqrt[3]{9x^2+4}}$$

$$Y = (9x^2+4)^{\frac{1}{3}} \leftarrow \textit{Function is rewritten}$$

$$Y' = \frac{1}{3}(9x^2+4)^{-\frac{2}{3}}\frac{d}{dx}(9x^2+4)$$

$$= \frac{1}{3}(9x^2+4)^{-\frac{2}{3}}(18x)$$

$$\boxed{Y' = \frac{6x}{\sqrt[3]{(9x^2+4)^2}}}$$

$$g(x) = \sqrt{x^2 - 2x + 1}$$

101

As $x^2 - 2x + 1$ it's a perfect square trinomial

$$g(x) = \sqrt{(x-1)^2}$$

By definition of absolute value

$$|x - 1| = \sqrt{(x-1)^2}$$

Now, by definition of absolute value:

Formally, the absolute value or module of any real number is defined by

$$|a| = \begin{cases} a, & \text{si } a \geq 0 \\ -a, & \text{si } a < 0 \end{cases}$$

$$g(x) = |x - 1|$$

Then you have to $\quad g(x) = x - 1$

$$g'(x) = 1$$

And you also have

$$g(x) = -(x - 1) = -x + 1$$

$$g'(x) = -1$$

$$\boxed{g'(x) = \begin{cases} -1, & x < 1 \\ 1, & x \geq 1 \end{cases}}$$

102

$$Y = 2\sqrt[4]{4 - x^2}$$

$$Y = 2(4 - x^2)^{\frac{1}{4}} \quad \leftarrow \textit{Function is rewritten}$$

$$Y'(x) = \frac{2}{4}(4 - x^2)^{-\frac{3}{4}} \frac{d}{dx}(4 - x^2)$$

$$= \frac{1}{2}(4 - x^2)^{-\frac{3}{4}}(-2x)$$

$$= \frac{-2x}{2\sqrt[4]{(4-x^2)^3}}$$

$$= \frac{-x}{\sqrt[4]{(4-x^2)^3}} \qquad \boxed{Y' = -\frac{x}{\sqrt[4]{(4 - x^2)^3}}}$$

103

$$f(x) = -3\sqrt[4]{2 - 9x}$$

$$f(x) = -3(2 - 9x)^{\frac{1}{4}} \leftarrow \textit{Function is rewritten}$$

$$f'(x) = -\frac{3}{4}(2 - 9x)^{-\frac{3}{4}} \frac{d}{dx}(2 - 9x)$$

$$= -\frac{3}{4}(2 - 9x)^{-\frac{3}{4}}(-9)$$

$$= \frac{27}{4\sqrt[4]{(2-9x)^3}}$$

$$\boxed{f'(x) = \frac{27}{4\sqrt[4]{(2 - 9x)^3}}}$$

$$Y = \frac{1}{(x-2)}$$

104

$$y = 1(x - 2)^{-1} \leftarrow \textit{Function is rewritten}$$

$$Y' = -1(x - 2)^{-2}\frac{d}{dx}(x - 2)$$

$$= -(x - 2)^{-2}(1)$$

$$= -\frac{1}{(x-2)^2}$$

$$Y' = -\frac{1}{(x-2)^2} \quad \boxed{\boldsymbol{Y' = -\frac{1}{(x-2)^2}}}$$

$$s(t) = \frac{1}{t^2 + 3t - 1}$$

105

$$s(t) = (t^2 + 3t - 1)^{-1} \leftarrow \textit{Function is rewritten}$$

$$s'(t) = -1(t^2 + 3t - 1)^{-2}\frac{d}{dt}(t^2 + 3t - 1)$$

$$= -(t^2 + 3t - 1)^{-2}(2t + 3)$$

$$= -\frac{2t+3}{(t^2+3t-1)^2}$$

$$\boxed{\boldsymbol{s'(t) = -\frac{2t + 3}{(t^2 + 3t - 1)^2}}}$$

106

$$f(t) = \left(\frac{1}{t-3}\right)^2$$

$$f(t) = \left(\frac{1}{t-3}\right)^2 \leftarrow \text{Function is rewritten}$$

$$f(t) = [(t-3)^{-1}]^2$$

$$f(t) = (t-3)^{-2}$$

$$f'(t) = -2(t-3)^{-3}\frac{d}{dt}(t-3)$$

$$= -2(t-3)^{-3}(1)$$

$$= -2(t-3)^{-3}$$

$$= -\frac{2}{(t-3)^3} \quad \boxed{f'(t) = -\frac{2}{(t-3)^3}}$$

107

$$Y = -\frac{5}{(t+3)^3}$$

$$Y = -5(t+3)^{-3} \leftarrow \text{Function is rewritten}$$

$$Y' = 15(t+3)^{-4}\frac{d}{dt}(t+3)$$

$$= 15(t+3)^{-4}(1)$$

$$= 15(t+3)^{-4}$$

$$\boxed{Y' = \frac{15}{(t+3)^4}}$$

$$Y = \frac{1}{\sqrt{x+2}}$$

108

$$Y = \frac{1}{(x+2)^{\frac{1}{2}}} \quad \leftarrow Function\ is\ rewritten$$

$$Y = 1(x+2)^{-\frac{1}{2}}$$

$$Y' = -\frac{1}{2}(x+2)^{-\frac{3}{2}}\frac{d}{dx}(x+2)$$

$$= -\frac{1}{2}(x+2)^{-\frac{3}{2}}(1)$$

$$= -\frac{1}{2}(x+2)^{-\frac{3}{2}}$$

$$\boxed{Y' = -\frac{1}{2\sqrt{x^3}}}$$

$$g(t) = \sqrt{\frac{1}{t^2-2}}$$

109

$$g(t) = \left[(t^2-2)^{-1}\right]^{\frac{1}{2}} \quad \leftarrow Function\ is\ rewritten$$

$$g(t) = (t^2-2)^{-\frac{1}{2}} \quad \rightarrow \quad g'(t) - \frac{1}{2}(t^2-2)^{-\frac{3}{2}}\frac{d}{dt}(t^2-2)$$

$$= -\frac{1}{2}(t^2-2)^{-\frac{3}{2}}(2t)$$

$$\boxed{\boldsymbol{g'(t) = -\frac{t}{\sqrt{(t^2-2)^3}}}}$$

110

$$f(x) = x^2 (x - 2)^4$$

The function is rewritten to have a better view of the product

$$f(x) = x^2[(x - 2)^4]$$

$$f'(x) = x^2 \frac{d}{dx}[(x - 2)^4] + [(x - 2)^4] \frac{d}{dx} x^2$$

$$= x^2[4(x - 2)^3] \frac{d}{dx}(x - 2) + [(x - 2)^4] 2x \frac{d}{dx}(x)$$

$$= x^2[4(x - 2)^3](1) + [(x - 2)^4] 2x (1)$$

$$= x^2[4(x - 2)^3] + [(x - 2)^4] 2x$$

$$= 4x^2(x - 2)^3 + 2x(x - 2)^4$$

$$= 2x(x - 2)^3[2x + (x - 2)] \leftarrow \textit{Common factor is taken out } 2x(x - 2)^3$$

$$= 2x(x - 2)^3[2x + x - 2]$$

$$= 2x(x - 2)^3[3x - 2]$$

$$= 2x(x - 2)^3(3x - 2)$$

$$\boxed{f'(x) = 2x(x - 2)^3(3x - 2)}$$

$$f(x) = x\,(3x - 9)^3$$

111

Analogously to the previous case, the function is rewritten

$$f(x) = x[(3x - 9)^3]$$

$$f'(x) = x\frac{d}{dx}[3x - 9)^3] + [3x - 9)^3]\frac{d}{dx}(x)$$

$$= x[3(3x - 9)^2]\frac{d}{dx}(3x - 9) + [(3x - 9)^3]\,(1)$$

$$= x[3(3x - 9)^2](3) + [(3x - 9)^3]$$

$$= 3x[3(3x - 9)^2] + [(3x - 9)^3]$$

$$= 9x(3x - 9)^2 + (3x - 9)^3$$

$$= (3x - 9)^2[9x + (3x - 9)] \leftarrow \textit{Common factor is taken out } (3x - 9)^2$$

$$= (3x - 9)^2[9x + 3x - 9]$$

$$= (3x - 9)^2[12x - 9]$$

$$= 3(3x - 9)^2(4x - 3)$$

$$= 3(9x^2 - 54x + 81)(4x - 3)$$

$$= 3[9(x^2 - 6x + 9)](4x - 3)$$

$$= 27(x - 3)^2(4x - 3)$$

$$\boxed{f'(x) = 27(x - 3)^2(4x - 3)}$$

112

$$Y(x) = x \sqrt{1 - x^2}$$

The function is rewritten

$$Y(x) = x \, (1 - x^2)^{\frac{1}{2}}$$

$$Y'(x) = x \frac{d}{dx} (1 - x^2)^{\frac{1}{2}} + (1 - x^2)^{\frac{1}{2}} \frac{d}{dx} (x)$$

$$= x \left[\frac{1}{2} (1 - x^2)^{-\frac{1}{2}} \right] \frac{d}{dx} (1 - x^2) + (1 - x^2)^{\frac{1}{2}} (1)$$

$$= x \left[\frac{1}{2} (1 - x^2)^{-\frac{1}{2}} \right] (-2x) + (1 - x^2)^{\frac{1}{2}}$$

$$= -x^2 (1 - x^2)^{-\frac{1}{2}} + (1 - x^2)^{\frac{1}{2}}$$

$$= (1 - x^2)^{-\frac{1}{2}} [-x^2 + (1 - x^2)] \quad \leftarrow Common\ factor\ is\ taken\ out (1 - x^2)^{-\frac{1}{2}}$$

$$= (1 - x^2)^{-\frac{1}{2}} [-x^2 + 1 - x^2]$$

$$= (1 - x^2)^{-\frac{1}{2}} [1 - 2x^2]$$

$$= \frac{1-2x^2}{(1-x^2)^{\frac{1}{2}}} \quad \boxed{Y'(x) = \frac{1 - 2x^2}{\sqrt{1 - x^2}}}$$

$$Y(x) = \frac{1}{2}x^2\sqrt{16 - x^2}$$

113

$$Y(x) = \frac{1}{2}x^2(16 - x^2)^{\frac{1}{2}} \quad \leftarrow \textit{The function is rewritten}$$

$$Y'(x) = \frac{1}{2}x^2 \frac{d}{dx}(16 - x^2)^{\frac{1}{2}} + (16 - x^2)^{\frac{1}{2}} \frac{d}{dx}\left(\frac{1}{2}x^2\right)$$

$$= \frac{1}{2}x^2 \left[\frac{1}{2}(16 - x^2)^{-\frac{1}{2}} \frac{d}{dx}(16 - x^2)\right] + (16 - x^2)^{\frac{1}{2}} (x)$$

$$= \frac{1}{2}x^2 \left[\frac{1}{2}(16 - x^2)^{-\frac{1}{2}} (-2x)\right] + (16 - x^2)^{\frac{1}{2}} (x)$$

$$= \frac{1}{2}x^2 \left[(16 - x^2)^{-\frac{1}{2}} (-x)\right] + (16 - x^2)^{\frac{1}{2}} (x)$$

$$= -\frac{1}{2}x^3 \left[(16 - x^2)^{-\frac{1}{2}}\right] + x (16 - x^2)^{\frac{1}{2}}$$

$$= x (16 - x^2)^{-\frac{1}{2}} \left[-\frac{1}{2}x^2 + (16 - x^2)\right]$$

$$= x (16 - x^2)^{-\frac{1}{2}} \left[-\frac{3}{2}x^2 + 16\right]$$

$$= \frac{x\left(\frac{-3x^2+32}{2}\right)}{(16-x^2)^{\frac{1}{2}}} \qquad \boxed{y'(x) = \frac{-x(3x^2 - 32)}{2\sqrt{16 - x^2}}}$$

114

$$y = \frac{x}{\sqrt{x^2 + 1}}$$

$$y = x\,(x^2 + 1)^{-\frac{1}{2}} \quad \leftarrow \textit{The function is rewritten}$$

$$Y' = x\frac{d}{dx}(x^2 + 1)^{-\frac{1}{2}} + (x^2 + 1)^{-\frac{1}{2}}\frac{d}{dx}(x)$$

$$= x\left(-\frac{1}{2}(x^2 + 1)^{-\frac{3}{2}}\frac{d}{dx}(x^2 + 1)\right) + (x^2 + 1)^{-\frac{1}{2}}(1)$$

$$= x\left(-\frac{1}{2}(x^2 + 1)^{-\frac{3}{2}}(2x)\right) + (x^2 + 1)^{-\frac{1}{2}}$$

$$= -x^2(x^2 + 1)^{-\frac{3}{2}} + (x^2 + 1)^{-\frac{1}{2}}$$

$$= (x^2 + 1)^{-\frac{3}{2}}\left[-x^2 + (x^2 + 1)\right]$$

$$= (x^2 + 1)^{-\frac{3}{2}}\left[-x^2 + x^2 + 1\right]$$

$$\boxed{Y' = \frac{1}{\sqrt{(x^2 + 1)^3}}}$$

APPLICATION OF THE RULE OF THE CHAIN TO TRIGONOMETRIC FUNCTIONS AND THEIR INVERSE, LOGARITHMIC AND EXPONENTIAL

$$Y = \cos(3x)$$

115

The chain rule continues to be applied

$$Y' = \frac{d}{dx}\cos(3x)\frac{d}{dx}(3x)$$

$$= -sen(3x)(3)$$

$$\boxed{Y' = -3sen(3x)}$$

$$Y = sen(\pi x)$$

116

$$Y' = \frac{d}{dx}sen(\pi x)\frac{d}{dx}(\pi x)$$

$$= \cos(\pi x)\pi$$

$$\boxed{Y' = \pi\cos(\pi x)}$$

$$g(x) = 3\tan(4x)$$

117

$$g'(x) = 3sec^2 4x\,(4)$$

$$\boxed{g'(x) = 12sec^2\,4x}$$

118

$$h(x) = \sec x^2$$

The function is rewritten as follows

$$h(x) = \sec (x^2)$$

$$h'(x) = \frac{d}{dx}\sec (x^2)\frac{d}{dx} (x^2)$$

$$= \sec(x^2)\tan(x^2)2x$$

$$\boxed{h'(x) = 2x \ \sec(x^2)\tan(x^2)}$$

119

$$Y = sen(\pi x)^2$$

$$Y' = \frac{d}{dx} sen(\pi x)^2 \frac{d}{dx}[(\pi x)^2]\frac{d}{dx} (\pi x) \ \leftarrow \textit{The chain rule is applied three times}$$

$$Y' = \cos(\pi x)^2(2\pi x)(\pi)$$

$$Y' = 2\pi^2 x \cos(\pi x)^2 \quad \boxed{Y' = 2\pi^2 x \cos(\pi x)^2}$$

120

$$Y = \cos(1 - 2x)^2$$

$$Y' = \frac{d}{dx}[\cos(1 - 2x)^2]\frac{d}{dx}[(1 - 2x)^2]\frac{d}{dx}(1 - 2x)$$

$$= [-sen(1 - 2x)^2][2(1 - 2x) \ ](-2)$$

$$= 4(1 - 2x)sen(1 - 2x)^2$$

$$\boxed{Y' = 4(1 - 2x)sen(1 - 2x)^2}$$

$$h(x) = sen(2x)cos(2x)$$

121

$$h'(x) = sen(2x)\frac{d}{dx}\cos(2x)\frac{d}{dx}(2x) + \cos(2x)\frac{d}{dx}sen(2x)\frac{d}{dx}(2x)$$

$$= sen(2x)(-sen2x(2)) + \cos(2x)cos(2x)(2)$$

$$= -2sen^2 2x + 2cos^2 2x$$

$$= 2cos^2 2x - 2sen^2 2x$$

$$= 2(cos^2 2x - sen^2 2x)$$

$$= 2(cos2\,x)$$

$$= 2cos(4x) \rightarrow \boxed{h'(x) = 2cos(4x)}$$

Note: $cos2u = cos^2 u - sen^2 u$. The angle of the problem is $2x$, when multiplying by 2 of the formula the result is $4x$.

$$g(\theta) = \sec\left(\frac{1}{2}\theta\right)\tan\left(\frac{1}{2}\theta\right)$$

122

$$g'(\theta) = \sec\left(\frac{1}{2}\theta\right)\frac{d}{d\theta}\tan\left(\frac{1}{2}\theta\right)\frac{d}{d\theta}\left(\frac{1}{2}\theta\right) + \tan\left(\frac{1}{2}\theta\right)\frac{d}{d\theta}\sec\left(\frac{1}{2}\theta\right)\frac{d}{d\theta}\left(\frac{1}{2}\theta\right)$$

$$= \sec\left(\frac{1}{2}\theta\right)sec^2\left(\frac{1}{2}\theta\right)\left(\frac{1}{2}\right) + \tan\left(\frac{1}{2}\theta\right)\sec\left(\frac{1}{2}\theta\right)\tan\left(\frac{1}{2}\theta\right)\left(\frac{1}{2}\right)$$

$$= \frac{1}{2}sec^3\frac{1}{2}\theta + \frac{1}{2}sec\left(\frac{1}{2}\theta\right)tan^2\frac{1}{2}\theta$$

$$= \frac{1}{2}sec\frac{1}{2}\theta\left[sec^2\frac{1}{2}\theta + tan^2\frac{1}{2}\theta\right]$$

$$\boxed{g'(\theta) = \frac{1}{2}sec\frac{1}{2}\theta\left[sec^2\frac{1}{2}\theta + tan^2\frac{1}{2}\theta\right]}$$

123

$$f(x) = \frac{\cot x}{\operatorname{sen} x}$$

The quotient differentiation rule applies

$$f'(x) = \frac{\operatorname{sen} x \frac{d}{dx}(\cot x) - \cot x \frac{d}{dx}(\operatorname{sen} x)}{(\operatorname{sen} x)^2}$$

$$= \frac{\operatorname{sen} x(-\csc^2 x) - \cot x (\cos x)}{\operatorname{sen}^2 x}$$

$$= \frac{-\operatorname{sen} x \csc^2 x - \cot x \cos x}{\operatorname{sen}^2 x}$$

$$= \frac{-\operatorname{sen} x \frac{1}{\operatorname{sen}^2 x} - \frac{\cos x}{\operatorname{sen} x} \cos x}{\operatorname{sen}^2 x}$$

$$= \frac{-\frac{1}{\operatorname{sen} x} - \frac{\cos^2 x}{\operatorname{sen} x}}{\operatorname{sen}^2 x}$$

$$= \frac{\frac{-1 - \cos^2 x}{\operatorname{sen} x}}{\operatorname{sen}^2 x}$$

$$= \frac{-1 - \cos^2 x}{\operatorname{sen}^3 x}$$

$$\boxed{f'(x) = -\frac{(1 + \cos^2 x)}{\operatorname{sen}^3 x}}$$

$$h(t) = 2cot^2(\pi t + 2)$$

124

$$h(t) = 2(\cot(\pi t + 2))^2 \leftarrow \textit{The function is rewritten}$$

$$h'(t) = 2\frac{d}{dt}[(\cot(\pi t + 2))^2]$$

$$= 2[2(\cot(\pi t + 2))]\frac{d}{dt}(\cot(\pi t + 2)\frac{d}{dt}(\pi t + 2)$$

$$= 2[2(\cot(\pi t + 2))](-csc^2(\pi t + 2)(\pi)$$

$$= -4\pi \cot(\pi t + 2)\, csc^2(\pi t + 2)$$

$$\boxed{h'(t) = -4\pi \cot(\pi t + 2)\, csc^2(\pi t + 2)}$$

$$Y = 4\ sec^2 x$$

125

$$Y = 4(secx)^2 \leftarrow \textit{The function is rewritten}$$

$$Y' = 4\frac{d}{dx}(\sec x)^2$$

$$= 8(\sec x)\frac{d}{dx}(\sec x)$$

$$= 8\sec x \sec x \tan x$$

$$= 8sec^2 x \tan x$$

$$= 8\frac{1}{cos^2 x}\frac{sen\, x}{\cos x}$$

$$\boxed{Y' = \frac{8\, sen\, x}{cos^3 x}}$$

126

$$f(x) = arcsen\ x^{cos^2 x}$$

For the resolution of the problem requires the use of the following formula

$$f(x) = arc\ sen\ u \quad and\ its\ derivative\ is: f'(x) = \frac{u'}{\sqrt{1-u^2}}$$

Then, to find the derivative it is required u'

$$Sea\ u = x^{(\cos x)^2}$$

We proceed to apply Neperian to both members of the previous equation

$$\ln u = (\cos x)^2 lnx$$

Next, both members of the equation are derived

$$\frac{u'}{u} = \left[2\cos x\ (-sen\ x)\ln x + (\cos x)^2\ \frac{1}{x}\right]$$

$$u' = \left[2\cos x\ (-sen\ x)\ln x + (\cos x)^2\ \frac{1}{x}\right]u$$

$$u' = \left[2\cos x\ (-sen\ x)\ln x + (\cos x)^2\ \frac{1}{x}\right]x^{(\cos x)^2}$$

Where u' represents the numerator in the formula $f'(x) = \dfrac{u'}{\sqrt{1-u^2}}$

Therefore, the derivative is

$$f'(x) = \frac{\left[-2sen\ x\cos x\ \ln x + \dfrac{cos^2 x}{x}\right]x^{cos^2 x}}{\sqrt{1-(x^{cos^2 x})^2}}$$

$$y = \frac{arcsen\ x^2 - 3}{sen\ x}$$

127

Sea $u = x^2 \qquad u' = 2x$

$$y' = \frac{sen\ x \frac{d}{dx}[arcsenx^2 - 3] - [arcsenx^2 - 3]\frac{d}{dx}(sen\ x)}{(sen\ x)^2}$$

$$y' = \frac{sen\ x \frac{2x}{\sqrt{1-x^4}} - [arcsen\ x^2 - 3]\cos x}{sen^2 x}$$

$$y' = \frac{\frac{2xsen\ x}{\sqrt{1-x^4}} - [arcsen\ x^2 - 3]\cos x}{sen^2 x}$$

$$\boxed{y' = \frac{\frac{2xsen\ x}{\sqrt{1-x^4}} - [arcsen\ x^2 - 3]\cos x}{sen^2 x}}$$

$$g(x) = 3arccos\frac{x}{2}$$

128

Sea $u = \frac{x}{2} \qquad u' = \frac{1}{2}$

For the resolution of the problem the use of the following formula is required

$$g(x) = arc\ cos\ u \ and\ its\ derivative\ is: g'(x) = \frac{-u'}{\sqrt{1-u^2}}$$

$$g' = \frac{-3*\frac{1}{2}}{\sqrt{1-\left(\frac{x}{2}\right)^2}} = \frac{-\frac{3}{2}}{\sqrt{1-\frac{x^2}{4}}} = \frac{-\frac{3}{2}}{\sqrt{\frac{4-x^2}{4}}} = \frac{-\frac{3}{2}}{\frac{1}{2}\sqrt{4-x^2}}$$

$$g' = \frac{-3}{\sqrt{4-x^2}} \quad \rightarrow \quad \boxed{g' = \frac{-3}{\sqrt{4-x^2}}}$$

129

$$y = \ln(t^2 + 4) - \frac{1}{2}\arctan\frac{t}{2}$$

For the resolution of the problem requires the use of the following formula

$$y = arctan\ u\ and\ its\ derivative\ is :\ y' = \frac{u'}{1+u^2}$$

To derive the tangent arc make u $= \dfrac{t}{2}$ $\rightarrow$ u′ $= \dfrac{1}{2}$

Now we proceed to derive the function y

$$y' = \frac{2t}{t^2+4} - \frac{1}{2}\frac{\frac{1}{2}}{1+\left(\frac{t}{2}\right)^2}$$

$$= \frac{2t}{t^2+4} - \frac{1}{2}\frac{\frac{1}{2}}{1+\frac{t^2}{4}}$$

$$= \frac{2t}{t^2+4} - \frac{1}{2}\frac{\frac{1}{2}}{\frac{t^2+4}{4}}$$

$$= \frac{2t}{t^2+4} - \frac{1}{2}\frac{4}{2(t^2+4)}$$

$$= \frac{2t}{t^2+4} - \frac{4}{4(t^2+4)}$$

$$= \frac{2t}{t^2+4} - \frac{1}{t^2+4}$$

$$\boxed{y' = \frac{2t-1}{t^2+4}}$$

$$y = arc\ cos\ (1 - 2\,x^2)$$

For the resolution of the problem requires the use of the following formula

$$y = arc\ cos\ u\ and\ its\ derivative\ is:\ y' = \frac{-u'}{\sqrt{1-u^2}}$$

Sea $u = 1 - 2\,x^2$ $\qquad\qquad u' = -4x$

$$y' = \frac{-(-4x)}{\sqrt{1-(1-2x^2)^2}}$$

$$= \frac{4x}{\sqrt{1-(1-4x^2+4x^4)}}$$

$$= \frac{4x}{\sqrt{1-1+4x^2-4x^4}}$$

$$= \frac{4x}{\sqrt{4x^2-4x^4}}$$

$$= \frac{4x}{\sqrt{4x^2(1-x^2)}}$$

$$= \frac{4x}{2x\sqrt{(1-x^2)}}$$

$$= \frac{2}{\sqrt{(1-x^2)}}$$

$$\boxed{y' = \frac{2}{\sqrt{1 - x^2}}}$$

131

$$y = arctan\ \frac{1+x}{1-x}$$

For the resolution of the problem requires the use of the following formula

$$y = arc\ tan\ u\ and\ its\ derivative\ is:\ y\ ' = \frac{u'}{1+u^2}$$

$$Sea\ \ u = \frac{1+x}{1-x}\ ,\ \ u' = \frac{(1-x)-(1+x)(-1)}{(1-x)^2} = \frac{1-x+1+x}{(1-x)^2}$$

$$u' = \frac{2}{(1-x)^2}$$

$$y' = \frac{\dfrac{2}{(1-x)^2}}{1+\dfrac{(1+x)^2}{(1-x)^2}}$$

$$= \frac{\dfrac{2}{(1-x)^2}}{\dfrac{(1-x)^2+(1+x)^2}{(1-x)^2}}$$

$$= \frac{2}{(1-x)^2+(1+x)^2}$$

$$= \frac{2}{1-2x+x^2+1+2x+x^2}$$

$$= \frac{2}{2x^2+2}$$

$$= \frac{1}{x^2+1}$$

$$\boxed{y' = \frac{1}{1+x^2}}$$

$Y = \sqrt{x}\, e^x$

132

$$Y = x^{\frac{1}{2}}\, e^x \leftarrow \textit{Function is rewritten}$$

$$y' = x^{\frac{1}{2}}\, \frac{d}{dx}(e^x) + e^x \frac{d}{dx}\left(x^{\frac{1}{2}}\right)$$

$$= x^{\frac{1}{2}}(e^x)\frac{dx}{dx} + e^x\left(\frac{1}{2}x^{-\frac{1}{2}}\right)$$

$$= \sqrt{x}\, e^x + \frac{e^x}{2\sqrt{x}}$$

$$= \frac{2\sqrt{x}\sqrt{x}\, e^x + e^x}{2\sqrt{x}}$$

$$= \frac{2x\, e^x + e^x}{2\sqrt{x}} = \frac{e^x(2x+1)}{2\sqrt{x}} \qquad \boxed{y' = \frac{e^x(2x+1)}{2\sqrt{x}}}$$

$y = x\, e^x$

133

$$y' = x\frac{d}{dx}\, e^x + e^x \frac{d}{dx}(x)$$

$$= x\, e^x \frac{dx}{dx} + e^x(1) \quad \leftarrow \textit{Rule of the chain}$$

$$= x\, e^x + e^x = e^x(x+1)$$

$$\boxed{y' = e^x(x+1)}$$

134

$$y = \sqrt{x\,e^x}$$

$$y = (xe^x)^{\frac{1}{2}} \quad \leftarrow \textit{Function is rewritten}$$

$$y' = \frac{1}{2}(xe^x)^{-\frac{1}{2}}\frac{d}{dx}(xe^x) \quad \leftarrow \textit{Rule of the chain}$$

$$= \frac{1}{2}(xe^x)^{-\frac{1}{2}}(xe^x + e^x)$$

$$= \frac{e^x(x+1)}{2\sqrt{xe^x}}$$

$$\boxed{y' = \frac{e^x(x+1)}{2\sqrt{xe^x}}}$$

135

$$y = x^3 e^x$$

$$y' = x^3 \frac{d}{dx}e^x + e^x \frac{d}{dx}x^3 \quad \leftarrow \textit{Derivative of a product is applied}$$

$$y' = x^3 e^x + e^x(3x^2)$$

$$= x^3 e^x + 3x^2 e^x$$

$$= x^2 e^x(x+3) \quad \leftarrow \textit{Common factor is taken out } x^2 e^x$$

$$\boxed{y' = x^2 e^x(x+3)}$$

$$y = \frac{e^x}{x^2}$$

136

$y = e^x x^{-2}$ ← *Function is rewritten*

$y' = e^x \dfrac{d}{dx} x^{-2} + x^{-2} \dfrac{d}{dx} e^x$

$\quad = e^x(-2x^{-3}) + x^{-2} e^x$

$\quad = \dfrac{e^x}{-2x^3} + \dfrac{e^x}{x^2}$

$\quad = -\dfrac{e^x}{2x^3} + \dfrac{e^x}{x^2}$

$\quad = \dfrac{-x^2 e^x + 2x^3 e^x}{2x^5}$

$\quad = \dfrac{x^2 e^x(-1+2x)}{2x^5} \quad = \dfrac{e^x(2x-1)}{2x^3}$ $\boxed{y' = \dfrac{e^x(2x-1)}{2x^3}}$

$$y = \frac{1}{x} e^x$$

137

$y = x^{-1} e^x$ ← *Function is rewritten*

$y' = x^{-1} \dfrac{d}{dx} e^x + e^x \dfrac{d}{dx} x^{-1}$

$\quad = x^{-1} e^x + e^x(-x^{-2})$

$\quad = \dfrac{e^x}{x} - \dfrac{e^x}{x^2}$

$\quad = \dfrac{x e^x - e^x}{x^2}$

$\quad = \dfrac{e^x(x-1)}{x^2}$ $\boxed{y' = \dfrac{e^x(x-1)}{x^2}}$

138

$$y = e^{3x}$$

$$y' = \frac{d}{dx} e^{3x}$$

$$= e^{3x} \frac{d}{dx}(3x) \quad \leftarrow \text{Rule of the chain}$$

$$= e^{3x}(3)$$

$$= 3e^{3x}$$

$$\boxed{y' = 3e^{3x}}$$

139

$$y = \log_2\left(e^{5ln\sqrt{x^7-x^5+3}}\right)$$

$$y' = \frac{d}{dx}\left[\log_2\left(e^{5ln\sqrt{x^7-x^5+3}}\right)\right] \frac{d}{dx}\left(e^{5ln\sqrt{x^7-x^5+3}}\right) \quad \rightarrow$$

$$\rightarrow \frac{d}{dx}\left(5ln\sqrt{x^7-x^5+3}\right) \frac{d}{dx}\left(\sqrt{x^7-x^5+3}\right) \frac{d}{dx}(x^7-x^5+3)$$

$$y' = \frac{1}{e^{5ln\sqrt{x^7-x^5+3}}} \log_2 e \; e^{5ln\sqrt{x^7-x^5+3}} \quad \rightarrow$$

$$\rightarrow 5 \frac{1}{\sqrt{x^7-x^5+3}} \frac{1}{2\sqrt{x^7-x^5+3}} \; 7x^6 - 5x^4$$

$$y' = \log_2 \frac{5}{\sqrt{x^7-x^5+3}} \frac{7x^6-5x^4}{2\sqrt{x^7-x^5+3}}$$

$$\boxed{y' = \frac{5log_2 e\,(7x^6-5x^4)}{2(x^7-x^5+3)}}$$

Note: $y = \log_a x \rightarrow y' = \frac{1}{x} \log_a e$

$$y = sen\ (\log(-x))$$

140

$$y' = \frac{d}{dx}[sen\ (\log(-x))]\ \frac{d}{dx}(\log(-x))\ \frac{d}{dx}(-x)$$

$$= \cos(\log(-x))\ \frac{1}{-x}\ loge\ (-1)$$

$$= \frac{\cos(\log(-x))\ loge}{x}$$

$$\boxed{y' = \frac{\mathbf{\cos(\log(-x))}\ \boldsymbol{loge}}{\boldsymbol{x}}}$$

$$g(x) = \sqrt[3]{\ln(x + sen(x))}$$

141

$$g(x) = [\ln(x + sen(x))]^{\frac{1}{3}} \quad \leftarrow \textit{Function is rewritten}$$

$$g'(x) = \frac{1}{3}\left[\ln\big(x + sen(x)\big)\right]^{-\frac{2}{3}}\ \frac{d}{dx}(\ln\big(x + sen(x)\big)\frac{d}{dx}(x + sen(x))$$

$$= \frac{1}{3}\left[\ln\big(x + sen(x)\big)\right]^{-\frac{2}{3}}\ \frac{1}{(x+sen(x))}\ (1 + \cos(x))$$

$$= \frac{(1+\cos x)}{3(x+senx)\,[\ln(x+senx)]^{\frac{2}{3}}}$$

$$= \frac{(1+\cos x)}{3(x+senx)\sqrt[3]{\ln(x+senx)^2}}$$

$$\boxed{g'(x) = \frac{\mathbf{1 + \cos x}}{\mathbf{3(x + senx)\sqrt[3]{\ln(x + senx)^2}}}}$$

142

$$y = x \, lnx$$

$$y' = x \frac{d}{dx} lnx + lnx \frac{d}{dx}(x) \quad \leftarrow Derivative \ of \ a \ product \ is \ applied$$

$$y' = x \frac{1}{x} + lnx(1)$$

$$= 1 + lnx$$

$$\boxed{y' = lnx + 1}$$

143

$$y = \frac{lnx}{x}$$

$$y = x^{-1} lnx \quad \leftarrow Function \ is \ rewritten$$

$$y' = x^{-1} \frac{d}{dx} lnx + lnx \frac{d}{dx} x^{-1}$$

$$= (x^{-1}) \frac{1}{x} + lnx(-x^{-2})$$

$$= \frac{1}{x^2} - \frac{lnx}{x^2}$$

$$= \frac{1 - lnx}{x^2}$$

$$\boxed{y' = \frac{1 - lnx}{x^2}}$$

$$y = \sqrt{x}\ \ln x$$

144

$$y = x^{\frac{1}{2}}\ln x \quad \leftarrow \textit{Function is rewritten}$$

$$y' = x^{\frac{1}{2}}\frac{d}{dx}\ln x + \ln x\frac{d}{dx}x^{\frac{1}{2}}$$

$$= x^{\frac{1}{2}}\ \frac{1}{x} + \ln x\left(\frac{1}{2}x^{-\frac{1}{2}}\right)$$

$$= \frac{\sqrt{x}}{x} + \frac{\ln x}{2\sqrt{x}} \quad \leftarrow \textit{It proceeds to simplify}$$

$$= \frac{2\sqrt{x}\sqrt{x}+x\ln x}{2x\sqrt{x}}$$

$$= \frac{2x+x\ln x}{2x\sqrt{x}}$$

$$= \frac{x(2+\ln x)}{2x\sqrt{x}} \quad \leftarrow \textit{Common factor x is taken out}$$

$$= \frac{2+\ln x}{2\sqrt{x}}$$

$$\boxed{y' = \frac{2 + \ln x}{2\sqrt{x}}}$$

145

$$y = x^2 \ln x$$

$$y' = x^2 \frac{d}{dx} \ln x + \ln x \frac{d}{dx} x^2 \quad \leftarrow \textit{Derivative of a product is applied}$$

$$y' = x^2 \frac{1}{x} + \ln x \, (2x)$$

$$= x + 2x \ln x$$

$$= x \, (1 + 2\ln x) \quad \leftarrow \textit{Common factor } x \textit{ is taken out}$$

$$\boxed{y' = x \, (1 + 2\ln x)}$$

146

$$y = e^x \ln x$$

$$y' = e^x \frac{d}{dx} \ln x + \ln x \frac{d}{dx} e^x \quad \leftarrow \textit{Derivative of a product is applied}$$

$$= e^x \frac{1}{x} + \ln x (e^x)$$

$$= \frac{e^x}{x} + e^x \ln x$$

$$= e^x \left(\frac{1}{x} + \ln x \right) \quad \leftarrow \textit{Common factor is taken out } e^x$$

$$\boxed{y' = e^x \left(\frac{1}{x} + \ln x \right)}$$

$$y = \ln x^3$$

147

$$y' = \frac{d}{dx} \ln x^3 \quad \leftarrow \frac{d}{dx}[\ln x] = \frac{1}{x}\frac{dx}{dx}$$

$$y' = \frac{1}{x^3} \frac{d}{dx}[x^3]$$

$$= \frac{3x^2}{x^3} \quad \leftarrow \textit{Similar terms are simplified}$$

$$= \frac{3}{x} \quad \boxed{y' = \frac{3}{x}}$$

$$y = \ln^3 x$$

148

$$y = (\ln x)^3 \quad \leftarrow \textit{Function is rewritten}$$

$$y' = \frac{d}{dx}(\ln x)^3$$

$$= 3(\ln x)^2 \frac{d}{dx} \ln x \quad \leftarrow \textit{Rule of the chain}$$

$$= 3(\ln x)^2 \left(\frac{1}{x}\right) = \frac{3}{x} \ln^2 x$$

$$\boxed{y' = \frac{3}{x} \ln^2 x}$$

149

$$y = \ln^4 x^3$$

The function is rewritten as follows

$$y = (\ln x^3)^4$$

$$y' = 4(\ln x^3)^3 \frac{d}{dx} \ln x^3$$

$$= 4(\ln x^3)^3 \frac{1}{x^3} \frac{d}{dx}[x^3] \quad \leftarrow Rule\ of\ the\ chain$$

$$= 4(\ln x^3)^3 \frac{3x^2}{x^3} \quad \leftarrow Similar\ terms\ are\ simplified$$

$$= \frac{12(\ln x^3)^3}{x}$$

$$\boxed{y' = \frac{12\ln^3 x^3}{x}}$$

150

$$y = (e^x)^3$$

$$y' = 3(e^x)^2 \frac{d}{dx} e^x \quad \leftarrow Rule\ of\ the\ chain$$

$$y' = 3(e^x)^2 e^x$$

Power product of the same base $\rightarrow$ $y' = 3e^{2x} e^x = 3e^{3x}$

$$\boxed{y' = 3e^{3x}}$$

$$y = ln\sqrt{x^2}$$

151

The function is rewritten as follows

$$y = ln(x^2)^{\frac{1}{2}}$$

By potentiation property is obtained

$$y = ln\,x$$

Now we proceed to derive

$$y' = \frac{d}{dx}lnx \quad \leftarrow \quad \frac{d}{dx}[lnx] = \frac{1}{x}\frac{dx}{dx}$$

$$y' = \frac{1}{x}$$

$$\boxed{y' = \frac{1}{x}}$$

$$y = \ln(3x - 1)$$

152

$$y' = \frac{1}{3x-1}\,\frac{d}{dx}(3x - 1) \quad \leftarrow \quad \textit{Rule of the chain}$$

$$y' = \frac{3}{3x-1}$$

$$\boxed{y' = \frac{3}{3x - 1}}$$

153

$$y = \frac{1}{lnx}$$

The function is rewritten as follows

$$y = 1\, ln^{-1}x$$

$$y = ln^{-1}x$$

$$y = (lnx)^{-1} \quad \leftarrow Proceed\ to\ derive$$

$$y' = -1(lnx)^{-2} \frac{d}{dx} lnx \quad \leftarrow\ Rule\ of\ the\ chain$$

$$y' = -(lnx)^{-2}\, \frac{1}{x}$$

$$y' = -\frac{1}{(lnx)^2 x} \qquad \boxed{y' = -\frac{1}{x\, ln^2 x}}$$

154

$$y = ln\sqrt{2x+1}$$

Proceed to derive

$$y' = \frac{d}{dx}\left[ln\sqrt{2x+1}\right] \frac{d}{dx}\left[\sqrt{2x+1}\right] \frac{d}{dx}\left[2x+1\right]$$

$$y' = \frac{\frac{1}{2}(2x+1)^{-\frac{1}{2}}(2)}{\sqrt{2x+1}}$$

$$y' = \frac{(2x+1)^{-\frac{1}{2}}}{(2x+1)^{\frac{1}{2}}}$$

$$y' = (2x+1)^{-1} \qquad \boxed{y' = \frac{1}{(2x+1)}}$$

$$y = (tan^{-1}x)^{tanx}$$

The function is rewritten as follows

$$y = [(tanx)^{-1}]^{tanx}$$

$$y = (tanx)^{-tanx}$$

Neperian is applied to both members of the equation

$$lny = \ln(tanx)^{-tanx}$$

$$lny = -tanx \ln(tanx)$$

Then we proceed to derive both members of the equation

$$\frac{y'}{y} = -tanx \frac{sec^2x}{tanx} + \ln(tanx)(-sec^2x)$$

$$\frac{y'}{y} = -sec^2x - \ln(tanx)(sec^2x)$$

$$\frac{y'}{y} = -\frac{1}{cos^2x} - \ln(tanx)\frac{1}{cos^2x}$$

$$\frac{y'}{y} = -\frac{1}{cos^2x}(1 + \ln(tanx))$$

$$\frac{y'}{y} = -\frac{1+\ln(tanx)}{cos^2x}$$

$$y' = -\frac{1+\ln(tanx)}{cos^2x} y$$

$$y' = -\frac{1+\ln(tanx)}{cos^2x} (tanx)^{-tanx}$$

$$\boxed{y' = -\frac{1 + \ln(tanx)}{cos^2x} (tanx)^{-tanx}}$$

156

$$y = (e^x)^{e^x}$$

Neperian is applied to both members of the equation

$$lny = ln(e^x)^{e^x} \quad \rightarrow \quad lny = e^x \ln(e^x)$$

Then we proceed to derive both members of the equation

$$\frac{y'}{y} = e^x \frac{e^x}{e^x} + \ln(e^x)\, e^x$$

$$\frac{y'}{y} = e^x + e^x \ln(e^x)$$

$$\frac{y'}{y} = e^x (1 + lne^x)$$

$$y' = e^x (1 + ln\, e^x)y \quad \boxed{y' = e^x (1 + \textbf{\textit{ln}}\, \boldsymbol{e^x})(e^x)^{e^x}}$$

157

$$y = (x^x)^x$$

The function is rewritten as follows

$$y = (x^x)^x \quad \rightarrow \quad y = x^{x^2}$$

Neperian is applied to both members of the equation and is derived

$$lny = ln\, x^{x^2} \quad \rightarrow \quad lny = x^2 \ln x$$

$$\frac{y'}{y} = 2xlnx + x^2 \frac{1}{x} \quad \rightarrow \quad \frac{y'}{y} = x + 2x \ln x \quad \rightarrow \quad y' = (x + 2x \ln x)y$$

$$\boxed{\boldsymbol{y' = (x + 2x \ln x)\; x^{x^2}}}$$

$$y = (\sqrt{x})^{\sqrt{x}}$$

158

Neperian is applied to both members of the equation

$$\ln y = \ln (\sqrt{x})^{\sqrt{x}}$$

$$\ln y = \sqrt{x}\ \ln \sqrt{x}$$

Then proceed to derive both members of the equation

$$\frac{y'}{y} = \sqrt{x}\ \frac{\frac{1}{2\sqrt{x}}}{\sqrt{x}} + \ln \sqrt{x}\ \frac{1}{2\sqrt{x}}$$

$$\frac{y'}{y} = \frac{1}{2\sqrt{x}} + \frac{1}{2\sqrt{x}}\ \ln \sqrt{x}$$

$$\frac{y'}{y} = \frac{1}{2\sqrt{x}}\left(1 + \ln \sqrt{x}\right)$$

$$y' = \frac{1}{2\sqrt{x}}\left(1 + \ln \sqrt{x}\right)y$$

$$y' = \frac{1}{2\sqrt{x}}\left(1 + \ln \sqrt{x}\right)(\sqrt{x})^{\sqrt{x}}$$

$$\boxed{y' = \frac{\left(1 + \ln \sqrt{x}\right)(\sqrt{x})^{\sqrt{x}}}{2\sqrt{x}}}$$

159

$$y = (\ln x)^{\ln x}$$

Neperian is applied to both members of the equation

$$lny = ln(\ln x)^{\ln x}$$

$$lny = \ln x[\, ln(\ln x)]$$

Then proceed to derive both members of the equation

$$\frac{y'}{y} = \ln x \left[\frac{\frac{1}{x}}{\ln x}\right] + \ln(\ln x)\,\frac{1}{x}$$

$$\frac{y'}{y} = \ln x \left[\frac{1}{x \ln x}\right] + \frac{\ln(\ln x)}{x}$$

$$\frac{y'}{y} = \frac{\ln x}{x \ln x} + \frac{\ln(\ln x)}{x}$$

$$\frac{y'}{y} = \frac{\ln x + \ln x\,[\ln(\ln x)]}{x \ln x}$$

$$\frac{y'}{y} = \frac{\ln x(1 + \ln(\ln x))}{x \ln x}$$

$$\frac{y'}{y} = \frac{(1 + \ln(\ln x))}{x}$$

$$y' = \frac{(1 + \ln(\ln x))}{x}\, y$$

$$\boxed{y' = \frac{1 + \ln(\ln x)}{x}\,(\ln x)^{\ln x}}$$

$$y = (\cos x)^{e^x}$$

160

Neperian is applied to both members of the equation

$$\ln y = \ln(cosx)^{e^x}$$

$$\ln y = e^x \ln(\cos x)$$

Then proceed to derive both members of the equation

$$\frac{y'}{y} = e^x \left(\frac{-sen\,x}{\cos x}\right) + \ln(\cos x)e^x$$

$$\frac{y'}{y} = -e^x \tan x + \ln(\cos x)e^x$$

$$\frac{y'}{y} = e^x[-\tan x + \ln(\cos x)]$$

$$y' = e^x[-\tan x + \ln(\cos x)]y$$

$$y' = e^x[-\tan x + \ln(\cos x)](\dot{cosx})^{e^x}$$

$$\boxed{y' = e^x[-\tan x + \ln(\cos x)](cosx)^{e^x}}$$

$$h(x) = e^{xlnx}\,2^{-x}$$

161

$$h'(x) = e^{xlnx}\frac{d}{dx}2^{-x} + 2^{-x}\frac{d}{dx}e^{xlnx} \quad \leftarrow \textit{Derivative of a product is applied}$$

$$h'(x) = e^{xlnx}\,ln2\,2^{-x}\,(-1) + 2^{-x}e^{xlnx}\left(x\frac{1}{x} + \ln x\right)$$

$$h'(x) = -e^{xlnx}ln2\,2^{-x} + 2^{-x}e^{xlnx}(\ln x + 1)$$

$$= e^{xlnx}2^{-x}(-ln2 + \ln x + 1)$$

$$\boxed{h'(x) = e^{xlnx}2^{-x}(-ln2 + \ln x + 1)}$$

162

$$k(x) = sen^3(\cos 5x)$$

The function is rewritten as follows

$$k(x) = [sen\,(\cos 5x)]^3$$

$$k'(x) = 3\,[sen\,(\cos 5x)]^2\,\frac{d}{dx}[sen\,(\cos 5x)]\frac{d}{dx}(\cos 5x)\frac{d}{dx}\,(5x)$$

$$k'(x) = 3\,[sen\,(\cos 5x)]^2(\cos(\cos 5x))(-sen5x)(5)$$

$$k'(x) = -15[sen\,(\cos 5x)]^2(\cos(\cos 5x))\,sen5x$$

$$\boxed{k'(x) = -15[sen\,(\cos 5x)]^2\,sen(5x)\,(\cos(\cos 5x))}$$

163

$$g(x) = \sqrt{sen^{-1}(3x)}$$

The function is rewritten as follow

$$g(x) = [sen(3x)^{-1}]^{\frac{1}{2}} \rightarrow g(x) = [sen(3x)]^{-\frac{1}{2}}$$

$$g'(x) = -\frac{1}{2}\,[sen(3x)]^{-\frac{3}{2}}\,\frac{d}{dx}(sen(3x))\frac{d}{dx}(3x)$$

$$g'(x) = -\frac{1}{2}\,[sen(3x)]^{-\frac{3}{2}}\,(\cos(3x))(3)$$

$$\boxed{g'(x) = -\frac{3cos3x}{2\sqrt{sen^3 3x}}}$$

$$h(x) = \sqrt{\ln(sen\ x)}$$

164

$$h(x) = [\ln(sen\ x)]^{\frac{1}{2}}$$

$$h'(x) = \frac{1}{2}[\ln(sen\ x)]^{-\frac{1}{2}}\frac{d}{dx}[\ln(sen\ x)]\frac{d}{dx}(sen\ x)$$

$$= \frac{1}{2}[\ln(sen\ x)]^{-\frac{1}{2}}\frac{1}{sen\ x}(\cos x)$$

$$= \frac{1}{2}[\ln(sen\ x)]^{-\frac{1}{2}}cotg\ x$$

$$= \frac{cotg\ x}{2\sqrt{\ln(sen\ x)}} \qquad \boxed{h'(x) = \frac{cotg\ x}{2\sqrt{\ln(sen\ x)}}}$$

$$t(x) = log_2\sqrt[3]{\frac{3-x}{x+3}}$$

165

We proceed to apply the rule of the chain

$$t'(x) = \frac{d}{dx}\left(log_2\sqrt[3]{\frac{3-x}{x+3}}\right)\frac{d}{dx}\left(\sqrt[3]{\frac{3-x}{x+3}}\right)\frac{d}{dx}\left(\frac{3-x}{x+3}\right)$$

$$t'(x) = \frac{1}{\sqrt[3]{\frac{3-x}{x+3}}}\ log_2 e\ \frac{1}{3\sqrt[3]{\left(\frac{3-x}{x+3}\right)^2}}\ \frac{(x+3)(-1)-(3-x)(1)}{(x+3)^2}$$

$$t'(x) = \frac{log_2 e\ (-x-3-3+x)}{3\sqrt[3]{\left(\frac{3-x}{x+3}\right)^3}\ (x+3)^2}$$

$$t'(x) = \frac{-6log_2 e}{3\left(\frac{3-x}{x+3}\right)(x+3)^2}$$

$$t'(x) = \frac{-2log_2 e}{(3-x)(x+3)} = \frac{-(-2log_2 e)}{-[(3-x)(x+3)]}$$

$$t'(x) = \frac{2log_2 e}{(x-3)(x+3)} \quad \leftarrow \textit{so you get a difference of squares}$$

$$\boxed{t'(x) = \frac{2log_2 e}{x^2 - 9}}$$

Expressing the result is terms of Neperian logarithm

$$log_a N = \frac{1}{\ln(a)} \; lnN \;\; \rightarrow \;\; log_2 e = \frac{1}{\ln(2)} \; lne \;\; \rightarrow \;\; log_2 e = \frac{1}{\ln(2)}$$

$$\boxed{t'(x) = \frac{2}{\ln(2)(x^2 - 9)}}$$

---

166 $\boxed{y = \sqrt{1 + \sqrt{3x^2 - 4x}}}$

The function is rewritten as follow

$$y = \left(1 + \sqrt{3x^2 - 4x}\right)^{\frac{1}{2}}$$

$$y' = \frac{1}{2}\left(1 + \sqrt{3x^2 - 4x}\right)^{-\frac{1}{2}} \frac{d}{dx}\left(1 + \sqrt{3x^2 - 4x}\right)$$

$$= \frac{1}{2}\left(1 + \sqrt{3x^2 - 4x}\right)^{-\frac{1}{2}} \frac{1}{2\sqrt{3x^2-4x}}(6x - 4)$$

$$= \frac{6x-4}{4\sqrt{1+\sqrt{3x^2-4x}}\sqrt{3x^2-4x}}$$

$$= \frac{2(3x-2)}{4\sqrt{1+\sqrt{3x^2-4x}}\sqrt{3x^2-4x}}$$

$$= \frac{3x-2}{2\sqrt{1+\sqrt{3x^2-4x}}\sqrt{3x^2-4x}} \qquad \boxed{y' = \frac{3x - 2}{2\sqrt{1 + \sqrt{3x^2 - 4x}}\sqrt{3x^2 - 4x}}}$$

$$f(x) = \frac{1}{\sqrt{x}} e^{(x^2 - \cos x + lnx)}$$

167

The function is rewritten as follow

$$f(x) = e^{(x^2 - \cos x + lnx)} \, x^{-\frac{1}{2}}$$

$$f'(x) = e^{(x^2 - \cos x + lnx)} \frac{d}{dx}\left(x^{-\frac{1}{2}}\right) + x^{-\frac{1}{2}} \frac{d}{dx} \, e^{(x^2 - \cos x + lnx)}$$

$$= e^{(x^2 - \cos x + lnx)}\left(-\frac{1}{2}x^{-\frac{3}{2}}\right) + x^{-\frac{1}{2}} e^{(x^2 - \cos x + lnx)}\left(2x + sen\, x + \frac{1}{x}\right)$$

$$= -\frac{e^{(x^2 - \cos x + lnx)}}{2x^{\frac{3}{2}}} + \frac{\left(2x + sen\, x + \frac{1}{x}\right) e^{(x^2 - \cos x + lnx)}}{x^{\frac{1}{2}}}$$

$$= -\frac{e^{(x^2 - \cos x + lnx)}}{2\sqrt{x^3}} + \frac{\left(\frac{2x^2 + xsen\, x + 1}{x}\right) e^{(x^2 - \cos x + lnx)}}{\sqrt{x}}$$

$$= -\frac{e^{(x^2 - \cos x + lnx)}}{2\sqrt{x^3}} + \frac{\frac{(2x^2 + xsen\, x + 1) e^{(x^2 - \cos x + lnx)}}{x}}{\sqrt{x}}$$

$$= -\frac{e^{(x^2 - \cos x + lnx)}}{2\sqrt{x^3}} + \frac{(2x^2 + xsen\, x + 1) e^{(x^2 - \cos x + lnx)}}{x\sqrt{x}}$$

$$\boxed{f'(x) = -\frac{e^{(x^2 - \cos x + lnx)}}{2\sqrt{x^3}} + \frac{(2x^2 + xsen\, x + 1) e^{(x^2 - \cos x + lnx)}}{x\sqrt{x}}}$$

168

$$y = (4 - 2x^2)^3 e^{\sqrt{x}}$$

The function is rewritten as follow

$$y = (4 - 2x^2)^3 e^{x^{\frac{1}{2}}}$$

$$y' = (4 - 2x^2)^3 \frac{d}{dx} e^{x^{\frac{1}{2}}} + e^{x^{\frac{1}{2}}} \frac{d}{dx} (4 - 2x^2)^3$$

$$= (4 - 2x^2)^3 e^{x^{\frac{1}{2}}} \frac{d}{dx} \left(x^{\frac{1}{2}} \right) + e^{x^{\frac{1}{2}}} 3(4 - 2x^2)^2 \frac{d}{dx} (4 - 2x^2)$$

$$= (4 - 2x^2)^3 e^{x^{\frac{1}{2}}} \left(\frac{1}{2} x^{-\frac{1}{2}} \right) + e^{x^{\frac{1}{2}}} 3(4 - 2x^2)^2 (-4x)$$

$$= (4 - 2x^2)^2 e^{x^{\frac{1}{2}}} \left[(4 - 2x^2) \left(\frac{1}{2} x^{-\frac{1}{2}} \right) + (-12x) \right]$$

$$= (4 - 2x^2)^2 e^{x^{\frac{1}{2}}} \left[\frac{(4 - 2x^2)}{2\sqrt{x}} - 12x \right]$$

$$= (4 - 2x^2)^2 e^{\sqrt{x}} \left[\frac{4 - 2x^2 - 24x\sqrt{x}}{2\sqrt{x}} \right]$$

$$= (4 - 2x^2)^2 e^{\sqrt{x}} \left[\frac{2(2 - x^2 - 12x\sqrt{x})}{2\sqrt{x}} \right]$$

$$= (4 - 2x^2)^2 e^{\sqrt{x}} \left[\frac{(2 - x^2 - 12x\sqrt{x})}{\sqrt{x}} \right]$$

$$= (4 - 2x^2)^2 e^{\sqrt{x}} \left[\frac{-x^2 - 12x\sqrt{x} + 2}{\sqrt{x}} \right]$$

$$\boxed{y' = (4 - 2x^2)^2 e^{\sqrt{x}} \left[\frac{(-x^2 - 12x\sqrt{x} + 2)}{\sqrt{x}} \right]}$$

$$y = [x + (x + x^2)^{-3}]^{-5}$$

169

We apply the chain rule

$$y' = -5\left[x + (x + x^2)^{-3}\right]^{-6} \frac{d}{dx}[x + (x + x^2)^{-3}]$$

$$= -5[x + (x + x^2)^{-3}]^{-6}\left[1 - 3(x + x^2)^{-4}\frac{d}{dx}(x + x^2)\right]$$

$$= -5[x + (x + x^2)^{-3}]^{-6}\left[1 - 3(x + x^2)^{-4}(1 + 2x)\right]$$

$$= -\frac{5\left[1 - 3(x + x^2)^{-4}(1 + 2x)\right]}{\left[x + (x + x^2)^{-3}\right]^6}$$

$$\boxed{y' = -\frac{5[1 - 3(x + x^2)^{-4}(1 + 2x)]}{[x + (x + x^2)^{-3}]^6}}$$

$$f(x) = \left(\frac{3x^2 + 6x}{x^3 - 4}\right)^5$$

170

$$f'(x) = 5\left(\frac{3x^2 + 6x}{x^3 - 4}\right)^4 \frac{d}{dx}\left(\frac{3x^2 + 6x}{x^3 - 4}\right)$$

$$= 5\left(\frac{3x^2 + 6x}{x^3 - 4}\right)^4 \frac{(x^3 - 4)(6x + 6) - (3x^2 + 6x)(3x^2)}{(x^3 - 4)^2}$$

$$= 5\left(\frac{3x^2 + 6x}{x^3 - 4}\right)^4 \frac{6x^4 - 24x + 6x^3 - 24 - (9x^4 + 18x^3)}{(x^3 - 4)^2}$$

$$= 5 \left(\frac{3x^2 + 6x}{x^3 - 4} \right)^4 \frac{6x^4 - 24x + 6x^3 - 24 - 9x^4 - 18x^3}{(x^3 - 4)^2}$$

$$= 5 \left(\frac{3x^2 + 6x}{x^3 - 4} \right)^4 \frac{-3x^4 - 12x^3 - 24x - 24}{(x^3 - 4)^2}$$

$$\boxed{f'(x) = 5 \left(\frac{3x^2 + 6x}{x^3 - 4} \right)^4 \left[-\frac{3x^4 + 12x^3 + 24x + 24}{(x^3 - 4)^2} \right]}$$

171 — $\quad f(x) = \sqrt[3]{2e^x - 2^x + 1} + \ln^5 x$

The function is rewritten as follow

$$f(x) = (2e^x - 2^x + 1)^{\frac{1}{3}} + (\ln x)^5$$

$$f'(x) = \frac{1}{3}(2e^x - 2^x + 1)^{-\frac{2}{3}} \frac{d}{dx}(2e^x - 2^x + 1) + (5\ln x)^4 \frac{d}{dx}(\ln x)$$

$$= \frac{1}{3}(2e^x - 2^x + 1)^{-\frac{2}{3}} \left(2e^x - \ln2 \, 2^x(1) + (5\ln x)^4 \frac{1}{x} \right.$$

$$= \frac{1}{3}(2e^x - 2^x + 1)^{-\frac{2}{3}} (2e^x - \ln2 \, 2^x) + \frac{625\ln^4 x}{x}$$

$$= \frac{(2e^x - \ln2 \, 2^x)}{3(2e^x - 2^x + 1)^{\frac{2}{3}}} + \frac{625\ln^4 x}{x}$$

$$= \frac{(2e^x - 2^x \ln2)}{3(2e^x - 2^x + 1)^{\frac{2}{3}}} + \frac{625\ln^4 x}{x}$$

$$\boxed{f'(x) = \frac{(2e^x - 2^x \ln2)}{3\sqrt[3]{(2e^x - 2^x + 1)^2}} + \frac{625\ln^4 x}{x}}$$

IMPLICIT DERIVATION

In the previously solved problems the functions were explicitly enunciated. For example, in the equation:

$$y = 2x^2 - 6$$

The variable (y) is written explicitly as a function of x. However, some functions are only implicitly stated in an equation. For example,

$$x^2 - 2y^3 + 4y = 2$$

Then the following question arises: **How to find the dy/dx for the above equation where it is very difficult to clear (y) as an explicit function of x?**

In these types of situations, the so-called implicit derivation must be used. To be clear about this technique it is necessary to take into account that the derivation is made with respect to x. If there is a term to be derived where (and) it will be necessary to apply the chain rule.

Example:

$$\frac{d}{dx}[y^5] = 5y^4 \frac{dy}{dx}$$

$$x^2 + y^2 = 16$$

172

Note that the function is written implicitly. Therefore, to find the derivative it is necessary to apply the implicit derivation. Keep in mind that it is derived in terms of x. If a term appears y, the chain rule applies.

$$x^2 + y^2 = 16$$

$$2x + 2y\frac{dy}{dx} = 0 \rightarrow \textit{The chain rule is applied}$$

Then it clears $\dfrac{dy}{dx}$

$$2y\frac{dy}{dx} = -2x$$

$$\frac{dy}{dx} = \frac{-2x}{2y} = -\frac{x}{y} \quad \boxed{\frac{dy}{dx} = -\frac{x}{y}}$$

$$x^{\frac{1}{2}} + y^{\frac{1}{2}} = 9$$

173

$$\frac{1}{2}x^{-\frac{1}{2}} + \frac{1}{2}y^{-\frac{1}{2}}\frac{dy}{dx} = 0$$

$$\frac{1}{2\sqrt{x}} + \frac{1}{2\sqrt{y}}\frac{dy}{dx} = 0$$

$$\frac{1}{2\sqrt{y}}\frac{dy}{dx} = -\frac{1}{2\sqrt{x}}$$

$$\frac{dy}{dx} = -\frac{2\sqrt{y}}{2\sqrt{x}} = -\frac{\sqrt{y}}{\sqrt{x}} = -\sqrt{\frac{y}{x}} \quad \boxed{\frac{dy}{dx} = -\sqrt{\frac{y}{x}}}$$

174

$$x^3 - xy + y^2 = 4$$

The function is implicitly derived

$$3x^2 - \left(x\frac{dy}{dx} + y\right) + 2y\frac{dy}{dx} = 0 \quad \rightarrow \text{ The chain rule is applied twice}$$

$$3x^2 - x\frac{dy}{dx} - y + 2y\frac{dy}{dx} = 0$$

Group the $\dfrac{dy}{dx}$ on the left side of the equality and the other terms on the right side

$$-x\frac{dy}{dx} + 2y\frac{dy}{dx} = -3x^2 + y$$

It is factored $\dfrac{dy}{dx}$

$$\frac{dy}{dx}(-x + 2y) = -3x^2 + y$$

It clears $\dfrac{dy}{dx}$

$$\frac{dy}{dx} = \frac{-3x^2 + y}{-x + 2y}$$

$$\frac{dy}{dx} = \frac{y - 3x^2}{2y - x} \qquad \boxed{\frac{dy}{dx} = \frac{y - 3x^2}{2y - x}}$$

$$x^3 - 3x^2y + 2xy^2 = 12$$

175

The function is implicitly derived

$$3x^2 - \left(3x^2\frac{dy}{dx} + 6xy\right) + \left(2x\,2y\frac{dy}{dx} + 2y^2\right) = 0$$

Removing the parentheses you have

$$3x^2 - 3x^2\frac{dy}{dx} - 6xy + 2x\,2y\frac{dy}{dx} + 2y^2 = 0$$

Group the $\dfrac{dy}{dx}$ on the left side of the equality and the other terms on the right side

$$-3x^2\frac{dy}{dx} + 2x\,2y\frac{dy}{dx} = -3x^2 + 6xy - 2y^2$$

It is factored $\dfrac{dy}{dx}$

$$\frac{dy}{dx}\left(-3x^2 + 2x\,2y\right) = -3x^2 + 6xy - 2y^2$$

$$\frac{dy}{dx} = \frac{-3x^2 + 6xy - 2y^2}{(-3x^2 + 2x\,2y)}$$

$$\frac{dy}{dx} = \frac{-3x^2 + 6xy - 2y^2}{-3x^2 + 4xy}$$

$$\boxed{\frac{dy}{dx} = \frac{6xy - 2y^2 - 3x^2}{4xy - 3x^2}}$$

176

$$x^3 y^3 - y = x$$

The function is rescribed

$$x^3 y^3 - y - x = 0$$

The function is implicitly derived

$$x^3\, 3y^2\, \frac{dy}{dx} + y^3 3x^2 - \frac{dy}{dx} - 1 = 0$$

$$x^3\, 3y^2 \frac{dy}{dx} - \frac{dy}{dx} = -3x^2 y^3 + 1$$

$$\frac{dy}{dx}\left(3x^3 y^2 - 1\right) = -3x^2 y^3 + 1$$

$$\frac{dy}{dx} = \frac{-3x^2 y^3 + 1}{3x^3 y^2 - 1}$$

$$\boxed{\frac{dy}{dx} = \frac{1 - 3x^2 y^3}{3x^3 y^2 - 1}}$$

177

$$sen\, x + 2\cos 2y = 1$$

$$\cos x + \left(-2 sen 2y\, 2\frac{dy}{dx}\right) = 0$$

$$\cos x - 4 sen 2y \frac{dy}{dx} = 0$$

$$-4 sen 2y \frac{dy}{dx} = -\cos x$$

$$\boxed{\frac{dy}{dx} = \frac{\cos x}{4 sen(2y)}}$$

$$sen\ x = x(1 + \tan y)$$

178

$$sen\ x = x + x\tan y$$

$$sen\ x - x - x\tan y = 0$$

$$\cos x - 1 - \left(x\sec^2 y \frac{dy}{dx} + \tan y(1)\right) = 0$$

$$\cos x - 1 - x\sec^2 y \frac{dy}{dx} - \tan y = 0$$

$$-x\sec^2 y \frac{dy}{dx} = -\cos x + \tan y + 1$$

$$\frac{dy}{dx} = \frac{-\cos x + \tan y + 1}{-x\sec^2 y}$$

$$\boxed{\frac{dy}{dx} = \frac{\cos x - \tan y - 1}{x\sec^2 y}}$$

$$y = sen\ (xy)$$

179

$$\frac{dy}{dx} = \cos(xy) \frac{d}{dx}(xy)$$

$$\frac{dy}{dx} = \cos(xy)\left(x\frac{dy}{dx} + y\right)$$

$$\frac{dy}{dx} = x\cos(xy)\frac{dy}{dx} + y\cos(xy)$$

$$\frac{dy}{dx} - x\cos(xy)\frac{dy}{dx} = y\cos(xy)$$

$$\frac{dy}{dx}(1 - x\cos(xy)) = y\cos(xy) \qquad \boxed{\frac{dy}{dx} = \frac{y\cos(xy)}{1 - x\cos(xy)}}$$

180

$$x^2 - y^2 = 16$$

$$2x - 2y\frac{dy}{dx} = 0$$

$$-2y\frac{dy}{dx} = -2x$$

$$\frac{dy}{dx} = \frac{2x}{2y} \quad \boxed{\frac{dy}{dx} = \frac{x}{y}}$$

181

$$x^3 + y^3 = 8$$

$$3x^2 + 3y^2\frac{dy}{dx} = 0$$

$$3y^2\frac{dy}{dx} = -3x^2$$

$$\frac{dy}{dx} = \frac{-3x^2}{3y^2}$$

$$\frac{dy}{dx} = -\frac{x^2}{y^2} \quad \boxed{\frac{dy}{dx} = -\frac{x^2}{y^2}}$$

182

$$x^2y + y^2x = -2$$

$$x^2\frac{dy}{dx} + 2xy + y^2(1) + 2xy\frac{dy}{dx} = 0$$

$$x^2\frac{dy}{dx} + 2xy\frac{dy}{dx} = -2xy - y^2$$

$$\frac{dy}{dx}(x^2 + 2xy) = -2xy - y^2$$

$$\frac{dy}{dx} = \frac{-2xy-y^2}{\left(x^2+2xy\right)}$$

$$\frac{dy}{dx} = \frac{-y(y+2x)}{x(x+2y)} \qquad \boxed{\frac{dy}{dx} = -\frac{y(y+2x)}{x(x+2y)}}$$

$$2sen\,x\cos y = 1 \qquad\qquad 183$$

$$2sen\,x\left(-sen\,y\frac{dy}{dx}\right) + \cos y(2\cos x) = 0$$

$$-2sen\,x\,sen\,y\frac{dy}{dx} = -2\cos x\cos y$$

$$\frac{dy}{dx} = \frac{\cos x\cos y}{sen\,x\,sen\,y}$$

$$\frac{dy}{dx} = \cot x\,coty \qquad \boxed{\frac{dy}{dx} = \cot x\,coty}$$

$$\sqrt{xy} = x - 2y \qquad\qquad 184$$

$$(xy)^{\frac{1}{2}} - x + 2y = 0$$

$$\frac{1}{2}xy^{-\frac{1}{2}}\left(x\frac{dy}{dx} + y\right) - 1 + 2\frac{dy}{dx} = 0$$

$$\frac{1}{2\sqrt{xy}}\left(x\frac{dy}{dx} + y\right) - 1 + 2\frac{dy}{dx} = 0$$

$$\frac{x}{2\sqrt{xy}}\frac{dy}{dx} + 2\frac{dy}{dx} = 1 - \frac{y}{2\sqrt{xy}}$$

$$\frac{dy}{dx}\left(\frac{x}{2\sqrt{xy}} + 2\right) = 1 - \frac{y}{2\sqrt{xy}}$$

$$\frac{dy}{dx}\left(\frac{x+4\sqrt{xy}}{2\sqrt{xy}}\right) = \frac{2\sqrt{xy}-y}{2\sqrt{xy}}$$

$$\frac{dy}{dx} = \frac{\frac{2\sqrt{xy}-y}{2\sqrt{xy}}}{\left(\frac{x+4\sqrt{xy}}{2\sqrt{xy}}\right)}$$

$$\frac{dy}{dx} = \frac{2\sqrt{xy}-y}{x+4\sqrt{xy}}$$

$$\frac{dy}{dx} = \frac{2\sqrt{xy}-y}{4\sqrt{xy}+x} \qquad \boxed{\frac{dy}{dx} = \frac{2\sqrt{xy}-y}{4\sqrt{xy}+x}}$$

185 $\boxed{(sen\,\pi x + \cos\pi y)^2 = 2}$

$$2(sen\,\pi x + \cos\pi y)\left[\cos\pi x\,\pi + (-sen\,\pi y\,\pi\tfrac{dy}{dx})\right] = 0$$

$$2(sen\,\pi x + \cos\pi y)\left[\pi\cos\pi x - \pi sen\,\pi y\,\tfrac{dy}{dx}\right] = 0$$

$$\left[\pi\cos\pi x - \pi sen\,\pi y\,\tfrac{dy}{dx}\right] = \frac{0}{2(sen\,\pi x+\cos\pi y)}$$

$$\left[\pi\cos\pi x - \pi sen\,\pi y\,\tfrac{dy}{dx}\right] = 0$$

$$-\pi sen\,\pi y\,\frac{dy}{dx} = -\pi\cos\pi x$$

$$\boxed{\frac{dy}{dx} = \frac{\cos\pi x}{sen\,\pi y}}$$

$$\cot y = x - y$$

186

$$\cot y - x + y = 0$$

$$-\csc^2 y \frac{dy}{dx} - 1 + \frac{dy}{dx} = 0$$

$$-\csc^2 y \frac{dy}{dx} + \frac{dy}{dx} = 1$$

$$\frac{dy}{dx}\left(-\csc^2 y + 1\right) = 1$$

$$\frac{dy}{dx} = \frac{1}{-\csc^2 y + 1} \quad Como \ 1 + \cot^2 y = \csc^2 y \ \rightarrow \ 1 - \csc^2 y = -\cot^2 y$$

$$\frac{dy}{dx} = -\frac{1}{\cot^2 y} \ \rightarrow \ \frac{dy}{dx} = -\tan^2 y \quad \boxed{\frac{dy}{dx} = -tan^2 y}$$

$$x = sec\frac{1}{y}$$

187

$$1 = \sec\frac{1}{y}\tan\frac{1}{y}\left[\frac{-\frac{dy}{dx}}{y^2}\right]$$

$$1 = -\frac{\frac{dy}{dx}}{y^2}\sec\frac{1}{y}\tan\frac{1}{y}$$

$$-y^2 = \frac{dy}{dx}\sec\frac{1}{y}\tan\frac{1}{y}$$

$$\frac{dy}{dx} = -\frac{y^2}{\sec\frac{1}{y}\tan\frac{1}{y}} \quad \boxed{\frac{dy}{dx} = -y^2\cos\frac{1}{y}\cot\frac{1}{y}}$$

188
$$xy = 4$$

$$x\frac{dy}{dx} + y(1) = 0$$

$$x\frac{dy}{dx} = -y$$

$$\boxed{\frac{dy}{dx} = -\frac{y}{x}}$$

189
$$x^2 - y^3 = 0$$

$$x^2 - y^3 = 0$$

$$2x - 3y^2\frac{dy}{dx} = 0$$

$$-3y^2\frac{dy}{dx} = -2x$$

$$\frac{dy}{dx} = \frac{2x}{3y^2} \qquad \boxed{\frac{dy}{dx} = \frac{2}{3}\frac{x}{y^2}}$$

190
$$y^2 = \frac{x^2 - 4}{x^2 + 4}$$

$$2y\frac{dy}{dx} = \frac{(x^2+4)(2x) - (x^2-4)(2x)}{(x^2+4)^2}$$

$$2y\frac{dy}{dx} = \frac{2x^3 + 8x - (2x^3 - 8x)}{(x^2+4)^2}$$

$$2\,y\frac{dy}{dx} = \frac{2x^3+8x-2x^3+8x}{(x^2+4)^2}$$

$$2\,y\frac{dy}{dx} = \frac{16x}{(x^2+4)^2}$$

$$\frac{dy}{dx} = \frac{16x}{2y(x^2+4)^2} \qquad \boxed{\frac{dy}{dx} = \frac{8x}{y(x^2+4)^2}}$$

$$\boxed{(x+y)^3 = x^3 + y^3}$$

191

The formula of the cube of a sum is applied and similar terms are canceled

$$x^3 + 3x^2y + 3xy^2 + y^3 = x^3 + y^3$$

$$3x^2y + 3xy^2 = 0$$

Then we proceed to implicitly derive

$$3x^2\frac{dy}{dx} + 6xy + 6xy\frac{dy}{dx} + 3y^2 = 0$$

$$3x^2\frac{dy}{dx} + 6xy\frac{dy}{dx} = -3y^2 - 6xy$$

$$\frac{dy}{dx}\left(3x^2 + 6xy\right) = -\left(3y^2 + 6xy\right)$$

$$\frac{dy}{dx} = -\frac{3y^2+6xy}{3x^2+6xy}$$

$$\frac{dy}{dx} = -\frac{3y(y+2x)}{3x(x+2y)}$$

$$\frac{dy}{dx} = -\frac{y(y+2x)}{x(x+2y)} \qquad \boxed{\frac{dy}{dx} = -\frac{y(y+2x)}{x(x+2y)}}$$

192

$$x^{\frac{2}{3}} + y^{\frac{2}{3}} = 5$$

Then we proceed to implicitly derive

$$\frac{2}{3}x^{-\frac{1}{3}} + \frac{2}{3}y^{-\frac{1}{3}}\frac{dy}{dx} = 0$$

$$\frac{2}{3\sqrt[3]{x}} + \frac{2}{3\sqrt[3]{y}}\frac{dy}{dx} = 0$$

$$\frac{2}{3\sqrt[3]{y}}\frac{dy}{dx} = -\frac{2}{3\sqrt[3]{x}}$$

$$\frac{dy}{dx} = -\frac{\frac{2}{3\sqrt[3]{x}}}{\frac{2}{3\sqrt[3]{y}}}$$

$$\frac{dy}{dx} = -\frac{6\sqrt[3]{y}}{6\sqrt[3]{x}}$$

$$\frac{dy}{dx} = -\frac{\sqrt[3]{y}}{\sqrt[3]{x}} \quad \rightarrow \quad \frac{dy}{dx} = -\sqrt[3]{\frac{y}{x}} \quad \boxed{\frac{dy}{dx} = -\sqrt[3]{\frac{y}{x}}}$$

193

$$x^3 + y^3 = 4xy + 1$$

Then we proceed to implicitly derive

$$3x^2 + 3y^2\frac{dy}{dx} = 4\left(x\frac{dy}{dx} + y\right)$$

$$3y^2\frac{dy}{dx} - 4x\frac{dy}{dx} = -3x^2 + 4y$$

$$\frac{dy}{dx}(3y^2 - 4x) = 4y - 3x^2 \quad \boxed{\frac{dy}{dx} = \frac{4y - 3x^2}{3y^2 - 4x}}$$

$$xy^2 - x^2 + 4 = 0$$

194

Then we proceed to implicitly derive

$$2xy\frac{dy}{dx} + y^2 - 2x = 0$$

$$2xy\frac{dy}{dx} = 2x - y^2$$

$$\frac{dy}{dx} = \frac{2x-y^2}{2xy} \quad \boxed{\frac{dy}{dx} = \frac{2x - y^2}{2xy}}$$

$$\tan(x + y) = x$$

195

Then we proceed to implicitly derive

$$sec^2(x + y)\left(1 + \frac{dy}{dx}\right) = 1$$

Terms to clear are grouped $\frac{dy}{dx}$:

$$\left(1 + \frac{dy}{dx}\right) = \frac{1}{sec^2(x+y)}$$

$$\frac{dy}{dx} = \frac{1}{sec^2(x+y)} - 1$$

$$\frac{dy}{dx} = \frac{1-sec^2(x+y)}{sec^2(x+y)}$$

Como $1 + tan^2x = sec^2x$

$$\frac{dy}{dx} = \frac{-tan^2(x+y)}{sec^2(x+y)}$$

$$\frac{dy}{dx} = -\frac{sen^2(x+y)}{cos^2(x+y)}\frac{1}{sec^2(x+y)}$$

$$\frac{dy}{dx} = -\frac{sen^2(x+y)}{cos^2(x+y)} \, cos^2(x+y)$$

$$\frac{dy}{dx} = -sen^2(x+y) \quad \boxed{\frac{dy}{dx} = -sen^2(x+y)}$$

196 $\boxed{y^2 - 2y = x}$

Then we proceed to implicitly derive

$$2y\frac{dy}{dx} - 2\frac{dy}{dx} = 1$$

$$\frac{dy}{dx}(2y - 2) = 1$$

$$\frac{dy}{dx} = \frac{1}{2y-2} \quad \boxed{\frac{dy}{dx} = \frac{1}{2y-2}}$$

197 $\boxed{(y-1)^2 = 4(x+2)}$

$$(y-1)^2 = 4x + 8$$

$$2(y-1)\frac{dy}{dx} = 4$$

$$\frac{dy}{dx} = \frac{4}{2(y-1)}$$

$$\frac{dy}{dx} = \frac{2}{y-1} \quad \boxed{\frac{dy}{dx} = \frac{2}{y-1}}$$

$$x + xy - y^2 - 20 = 0$$

198

$$1 + x\frac{dy}{dx} + y - 2y\frac{dy}{dx} = 0$$

$$x\frac{dy}{dx} - 2y\frac{dy}{dx} = -y - 1$$

$$\frac{dy}{dx}(x - 2y) = -y - 1$$

$$\frac{dy}{dx} = \frac{-y-1}{x-2y}$$

The numerator and the denominator are multiplied by -1

$$\boxed{\frac{dy}{dx} = \frac{y + 1}{2y - x}}$$

$$x\cos y = 1$$

199

$$x\left(-sen\, y\frac{dy}{dx}\right) + \cos y = 0$$

$$-x\, sen\, y\frac{dy}{dx} = -\cos y$$

$$\frac{dy}{dx} = \frac{\cos y}{x\, sen\, y}$$

$$\frac{dy}{dx} = \frac{1}{x}\cot y \qquad \boxed{\frac{dy}{dx} = \frac{\cot y}{x}}$$

200

$$x^3 y^2 = 2x^2 + y^2$$

Then we proceed to implicitly derive

$$2x^3 y \frac{dy}{dx} + 3x^2 y^2 = 4x + 2y \frac{dy}{dx}$$

$$2x^3 y \frac{dy}{dx} - 2y \frac{dy}{dx} = 4x - 3x^2 y^2$$

$$\frac{dy}{dx}(2x^3 y - 2y) = 4x - 3x^2 y^2$$

$$\boxed{\frac{dy}{dx} = \frac{4x - 3x^2 y^2}{2x^3 y - 2y}} \quad \rightarrow \quad \boxed{\frac{dy}{dx} = \frac{x(4 - 3xy^2)}{y(2x^3 - 2)}}$$

201

$$x^5 - 6xy^3 + y^4 = 1$$

Then we proceed to implicitly derive

$$5x^4 - 18xy^2 \frac{dy}{dx} - 6y^3 + 4y^3 \frac{dy}{dx} = 0$$

$$-18xy^2 \frac{dy}{dx} + 4y^3 \frac{dy}{dx} = -5x^4 + 6y^3$$

$$\frac{dy}{dx}(4y^3 - 18xy^2) = 6y^3 - 5x^4$$

$$\boxed{\frac{dy}{dx} = \frac{6y^3 - 5x^4}{4y^3 - 18xy^2}}$$

$$y = (x - y)^2$$

Then we proceed to implicitly derive

$$\frac{dy}{dx} = 2(x - y)(1 - \frac{dy}{dx})$$

$$\frac{dy}{dx} = (2x - 2y)(1 - \frac{dy}{dx})$$

$$\frac{dy}{dx} = 2x - 2x\frac{dy}{dx} - 2y + 2y\frac{dy}{dx}$$

$$\frac{dy}{dx} + 2x\frac{dy}{dx} - 2y\frac{dy}{dx} = 2x - 2y$$

$$\frac{dy}{dx}(1 + 2x - 2y) = 2x - 2y$$

$$\frac{dy}{dx} = \frac{2x - 2y}{1 + 2x - 2y}$$

$$\frac{dy}{dx} = \frac{2x - 2y}{2x - 2y + 1}$$

$$\frac{dy}{dx} = \frac{2(x - y)}{2\left(x - y + \frac{1}{2}\right)}$$

$$\frac{dy}{dx} = \frac{(x - y)}{\left(x - y + \frac{1}{2}\right)} \quad \boxed{\frac{dy}{dx} = \frac{x - y}{x - y + \frac{1}{2}}}$$

203

$$y^4 - y^2 = 10x - 3$$

Then we proceed to implicitly derive

$$4y^3 \frac{dy}{dx} - 2y \frac{dy}{dx} = 10$$

$$\frac{dy}{dx}(4y^3 - 2y) = 10$$

$$\frac{dy}{dx} = \frac{10}{4y^3 - 2y}$$

$$\frac{dy}{dx} = \frac{10}{2y(2y^2 - 1)} \qquad \boxed{\frac{dy}{dx} = \frac{5}{y(2y^2 - 1)}}$$

204

$$(x - 1)^2 + (y + 4)^2 = 25$$

Then we proceed to implicitly derive

$$2(y + 4)\frac{dy}{dx} = -2(x - 1)$$

$$\frac{dy}{dx} = \frac{-2(x - 1)}{2(y + 4)}$$

$$\boxed{\frac{dy}{dx} = -\frac{x - 1}{y + 4}} \quad O \quad \boxed{\frac{dy}{dx} = \frac{1 - x}{y + 4}}$$

Pedro Pablo **CORONEL PÉREZ** / Pablo Josué **CORONEL LÓPEZ**
250 SOLVED EXERCISES FROM DERIVATIVES WITH APPLICATIONS

205

$$\frac{x+y}{x-y} = x$$

It derives both members of the equation

$$\frac{(x-y)\left(1+\frac{dy}{dx}\right)-(x+y)\left(1-\frac{dy}{dx}\right)}{(x-y)^2} = 1$$

$$\frac{x+x\frac{dy}{dx}-y-y\frac{dy}{dx}-\left(x-x\frac{dy}{dx}+y-y\frac{dy}{dx}\right)}{(x-y)^2} = 1$$

$$\frac{x+x\frac{dy}{dx}-y-y\frac{dy}{dx}-x+x\frac{dy}{dx}-y+y\frac{dy}{dx}}{(x-y)^2} = 1$$

$$x+x\frac{dy}{dx}-y-y\frac{dy}{dx}-x+x\frac{dy}{dx}-y+y\frac{dy}{dx} = (x-y)^2$$

Group the $\frac{dy}{dx}$ on the left side of the equality and the other terms on the right side

$$x\frac{dy}{dx}-y\frac{dy}{dx}+x\frac{dy}{dx}+y\frac{dy}{dx} = (x-y)^2-x+y+x+y$$

$$2x\frac{dy}{dx} = (x-y)^2+2y$$

$$\frac{dy}{dx} = \frac{(x-y)^2+2y}{2x}$$

$$\boxed{\frac{dy}{dx} = \frac{(x-y)^2+2y}{2x}}$$

206

$$y^2 = \frac{x-1}{x+2}$$

$$2y\frac{dy}{dx} = \frac{(x+2)(1)-(x-1)(1)}{(x+2)^2}$$

$$2y\frac{dy}{dx} = \frac{x+2-x+1}{(x+2)^2}$$

$$2y\frac{dy}{dx} = \frac{3}{(x+2)^2}$$

$$\boxed{\frac{dy}{dx} = \frac{3}{2y(x+2)^2}}$$

207

$$x = \sec y$$

$$1 = (\sec y \tan y)\frac{dy}{dx}$$

$$\frac{dy}{dx} = \frac{1}{\sec y \tan y}$$

$$\frac{dy}{dx} = \frac{1}{\frac{1}{\cos y}\frac{sen\,y}{\cos y}} = \frac{1}{\frac{1}{co\,y}\frac{sen\,y}{\cos y}}$$

$$\frac{dy}{dx} = \cos y\,\frac{\cos y}{sen\,y} \rightarrow \frac{dy}{dx} = \cos y \cot y$$

$$\boxed{\frac{dy}{dx} = \cos y \cot y}$$

$$x + y = \cos xy$$

208

It derives both members of the equation

$$1 + \frac{dy}{dx} = -sen\ xy\ (x\frac{dy}{dx} + y)$$

$$1 + \frac{dy}{dx} = -x\frac{dy}{dx}\ sen\ xy - ysen\ xy$$

Group the $\frac{dy}{dx}$ on the left side of the equality and the other terms on the right side

$$\frac{dy}{dx} + x\frac{dy}{dx}sen\ xy = -ysen\ xy - 1$$

$$\frac{dy}{dx}(1 + xsen\ xy) = -ysen\ xy - 1$$

$$\frac{dy}{dx} = \frac{-ysen\ xy-1}{1+xsen\ xy}$$

$$\boxed{\frac{dy}{dx} = \frac{-ysen\ xy - 1}{1 + xsen\ xy}}\quad \text{O} \quad \boxed{\frac{dy}{dx} = -\frac{ysen\ xy + 1}{1 + xsen\ xy}}$$

$$xy = sen(x + y)$$

209

$$x\frac{dy}{dx} + y = \cos(x + y)(1 + \frac{dy}{dx})$$

$$x\frac{dy}{dx} + y = \cos(x + y) + \cos(x + y)\frac{dy}{dx}$$

$$x\frac{dy}{dx} - \cos(x + y)\frac{dy}{dx} = \cos(x + y) - y$$

$$\frac{dy}{dx}[x - \cos(x + y)] = \cos(x + y) - y \qquad \boxed{\frac{dy}{dx} = \frac{\cos(x + y) - y}{x - \cos(x + y)}}$$

210

$$(x^2 + y^2)^6 = x^3 - y^3$$

$$6(x^2 + y^2)^5 \left(2x + 2y\frac{dy}{dx}\right) = 3x^2 - 3y^2\frac{dy}{dx}$$

$$12x(x^2 + y^2)^5 + 12y(x^2 + y^2)^5\frac{dy}{dx} = 3x^2 - 3y^2\frac{dy}{dx}$$

$$12y(x^2 + y^2)^5\frac{dy}{dx} + 3y^2\frac{dy}{dx} = 3x^2 - 12x(x^2 + y^2)^5$$

$$\frac{dy}{dx}\left[12y(x^2 + y^2)^5 + 3y^2\right] = 3x^2 - 12x(x^2 + y^2)^5$$

$$\frac{dy}{dx} = \frac{\left[3x^2 - 12x(x^2+y^2)^5\right]}{\left[12y(x^2+y^2)^5+3y^2\right]} \rightarrow \frac{dy}{dx} = \frac{3\left[x^2-4x(x^2+y^2)^5\right]}{3[4y(x^2+y^2)^5+y^2]}$$

$$\boxed{\frac{dy}{dx} = \frac{x^2 - 4x(x^2 + y^2)^5}{4y(x^2 + y^2)^5 + y^2}}$$

211

$$x^2 sen\,(x + y) - 5ye^x = 3$$

$$x^2 \cos(x + y)(1 + \frac{dy}{dx}) + sen\,(x + y)2x - 5\left(ye^x + e^x\frac{dy}{dx}\right) = 0$$

$$x^2 \cos(x + y)(1 + \frac{dy}{dx}) + 2xsen\,(x + y) - 5ye^x - 5e^x\frac{dy}{dx} = 0$$

$$x^2 \cos(x + y) + x^2 \cos(x + y)\frac{dy}{dx} + 2xsen\,(x + y) - 5ye^x - 5e^x\frac{dy}{dx} = 0$$

$$x^2 \cos(x + y)\frac{dy}{dx} - 5e^x\frac{dy}{dx} = -x^2 \cos(x + y) - 2xsen\,(x + y) + 5ye^x$$

$$\frac{dy}{dx}[x^2 \cos(x + y) - 5e^x] = -x^2 \cos(x + y) - 2xsen\,(x + y) + 5ye^x$$

$$\frac{dy}{dx} = \frac{-x^2 \cos(x+y) - 2xsen\,(x+y) + 5ye^x}{x^2 \cos(x+y) - 5e^x}$$

The numerator and the denominator are multiplied by -1

$$\frac{dy}{dx} = \frac{x^2 \cos(x+y) + 2x\,sen\,(x+y) - 5ye^x}{-x^2 \cos(x+y) + 5e^x}$$

$$x^2 \tan y = y\,sen\,x$$ **212**

$$x^2 sec^2 y \frac{dy}{dx} + 2x\tan y = y \cos x + sen\,x \frac{dy}{dx}$$

$$x^2 sec^2 y \frac{dy}{dx} - sen\,x \frac{dy}{dx} = -2x\tan y + y \cos x$$

$$\frac{dy}{dx}(x^2 sec^2 y - sen\,x) = y \cos x - 2x\tan y$$

$$\frac{dy}{dx} = \frac{y \cos x - 2x\tan y}{x^2 sec^2 y - sen\,x}$$

$$\tan y = 3x^2 + \tan(x+y)$$ **213**

$$sec^2 y \frac{dy}{dx} = 6x + sec^2(x+y)\left(1 + \frac{dy}{dx}\right)$$

$$sec^2 y \frac{dy}{dx} = 6x + sec^2(x+y) + sec^2(x+y)\frac{dy}{dx}$$

$$sec^2 y \frac{dy}{dx} - sec^2(x+y)\frac{dy}{dx} = 6x + sec^2(x+y)$$

$$\frac{dy}{dx}[sec^2 y - sec^2(x+y)] = 6x + sec^2(x+y)$$

$$\frac{dy}{dx} = \frac{6x + sec^2(x+y)}{sec^2 y - sec^2(x+y)}$$

214

$$\sqrt{\dfrac{y-\sqrt{x}}{y+\sqrt{x}}} + \sqrt{\dfrac{y+\sqrt{x}}{y-\sqrt{x}}} = \dfrac{5}{2}$$

The problem equation is written and simplified

$$\sqrt{\dfrac{y-\sqrt{x}}{y+\sqrt{x}}} + \sqrt{\dfrac{y+\sqrt{x}}{y-\sqrt{x}}} = \dfrac{\left(\sqrt{y-\sqrt{x}}\right)^2 + \left(\sqrt{y+\sqrt{x}}\right)^2}{\sqrt{y+\sqrt{x}}\,\sqrt{y-\sqrt{x}}} = \dfrac{5}{2}$$

$$= \dfrac{y-\sqrt{x}+y+\sqrt{x}}{\sqrt{(y+\sqrt{x})(y-\sqrt{x})}} = \dfrac{2y}{\sqrt{y^2-x}} = \dfrac{5}{2}$$

Then the equation to derive is $\quad \dfrac{2y}{\sqrt{y^2-x}} = \dfrac{5}{2}$

$$\dfrac{\sqrt{y^2-x}\left(2\dfrac{dy}{dx}\right) - 2y\left(\dfrac{2y\dfrac{dy}{dx}-1}{2\sqrt{y^2-x}}\right)}{\left(\sqrt{y^2-x}\right)^2} = 0$$

$$\sqrt{y^2-x}\left(2\dfrac{dy}{dx}\right) - y\left(\dfrac{2y\dfrac{dy}{dx}-1}{\sqrt{y^2-x}}\right) = 0$$

Eliminating denominators you get

$$2\dfrac{dy}{dx}\sqrt{y^2-x}\sqrt{y^2-x} - 2y^2\dfrac{dy}{dx} + y = 0$$

Mathematically operating the roots you have

$$2\dfrac{dy}{dx}\left(y^2-x\right) - 2y^2\dfrac{dy}{dx} + y = 0$$

$$2y^2\dfrac{dy}{dx} - 2x\dfrac{dy}{dx} - 2y^2\dfrac{dy}{dx} + y = 0$$

Canceling similar terms

$$-2x\frac{dy}{dx} + y = 0$$

$$-2x\frac{dy}{dx} = -y$$

$$\frac{dy}{dx} = \frac{-y}{-2x} \quad \boxed{\frac{dy}{dx} = \frac{y}{2x}}$$

$$\sec^2 y + \cot(x - y) = \tan^2 x$$

215

$$2\sec y(\sec y \tan y)\frac{dy}{dx} + \left[-\csc^2(x-y)(1 - \frac{dy}{dx})\right] = 2\tan x\, \sec^2 x$$

$$2\sec y(\sec y \tan y)\frac{dy}{dx} + \left[-\csc^2(x-y) + \csc^2(x-y)\frac{dy}{dx}\right] = 2\tan x\, \sec^2 x$$

$$2\sec y(\sec y \tan y)\frac{dy}{dx} - \csc^2(x-y) + \csc^2(x-y)\frac{dy}{dx} = 2\tan x\, \sec^2 x$$

$$2\sec y(\sec y \tan y)\frac{dy}{dx} + \csc^2(x-y)\frac{dy}{dx} = 2\tan x\, \sec^2 x + \csc^2(x-y)$$

$$\frac{dy}{dx}[2\sec y(\sec y \tan y) + \csc^2(x-y)] = 2\tan x\, \sec^2 x + \csc^2(x-y)$$

$$\frac{dy}{dx} = \frac{2\tan x \sec^2 x + \csc^2(x-y)}{2\sec y \sec y \tan y + \csc^2(x-y)} \rightarrow$$

$$\boxed{\frac{dy}{dx} = \frac{2\tan x\, \sec^2 x + \csc^2(x-y)}{2\sec^2 y\, \tan y + \csc^2(x-y)}}$$

216

$$y = \cos(x - y)$$

$$\frac{dy}{dx} = -sen\,(x - y)\left(1 - \frac{dy}{dx}\right)$$

$$\frac{dy}{dx} = -sen\,(x - y) + sen\,(x - y)\frac{dy}{dx}$$

$$\frac{dy}{dx} - sen\,(x - y)\frac{dy}{dx} = -sen\,(x - y)$$

$$\frac{dy}{dx}\left(1 - sen\,(x - y)\right) = -sen\,(x - y)$$

$$\frac{dy}{dx} = \frac{-sen\,(x-y)}{1-sen\,(x-y)} = \frac{sen\,(x-y)}{sen\,(x-y)-1}$$

$$\boxed{\frac{dy}{dx} = \frac{sen\,(x - y)}{sen\,(x - y) - 1}}$$

APPLY THE L'HOPITAL RULE TO EVALUATE A LIMIT

$$\lim_{x \to 0} \frac{sen\ 4x}{2x}$$

217

When evaluating the limit by direct substitution we find an indetermination of the form $\frac{0}{0}$. Therefore, the L'Hoital rule can be applied. That is, both the numerator and the denominator are derived.

$$\lim\nolimits_{x \to 0} \frac{sen\ 4x}{2x} = \frac{sen4(0)}{2(0)} = \frac{sen\ (0)}{0} = \frac{0}{0}$$

$$\lim\nolimits_{x \to 0} \frac{sen\ 4x}{2x} = \lim\nolimits_{x \to 0} \frac{\frac{d}{dx}[sen\ 4x]}{\frac{d}{dx}[2x]}$$

$$\lim\nolimits_{x \to 0} \frac{cos4x(4)}{2} = 2\lim_{x \to 0} cos4x$$

When evaluating the limit the result is:

$$2\lim_{x \to 0} cos4x = 2\ cos4(0) = 2\cos(0) = 2 * 1 = 2$$

As a result, you get
$$\lim_{x \to 0} \frac{sen\ 4x}{2x} = 2$$

$$\lim_{x \to \infty} \frac{2x + 1}{4x^2 + x}$$

218

When evaluating the limit by direct substitution we find an indetermination of the form $\frac{\infty}{\infty}$. Therefore, the L'Hoital rule can be applied. That is, both the numerator and the denominator are derived.

$$\lim\nolimits_{x \to 0} \frac{2x+1}{4x^2+x} = \frac{2(\infty)+1}{4\infty^2+\infty} = \frac{\infty}{\infty}$$

$$\lim_{x\to 0}\frac{2x+1}{4x^2+x} = \lim_{x\to 0}\frac{\frac{d}{dx}[2x+1]}{\frac{d}{dx}[4x^2+x]}$$

$$= \lim_{x\to 0}\frac{2}{8x+1}$$

When evaluating the limit the result is:

$$\lim_{x\to 0}\frac{2}{8x+1} = \frac{2}{8(\infty)+1} = \frac{2}{\infty} = 0$$

As a result, you get

$$\boxed{\lim_{x\to\infty}\frac{2x+1}{4x^2+x} = 0}$$

219 $\qquad\qquad\boxed{\lim_{x\to 0}\frac{sen\ 2x}{sen\ 3x}}$

When evaluating the limit by direct substitution we find an indetermination of the form $\frac{0}{0}$. Therefore, the L'Hoital rule can be applied. That is, both the numerator and the denominator are derived.

$$\lim_{x\to 0}\frac{sen\ 2x}{sen\ 3x} = \frac{sen\ 2(0)}{sen\ 3(0)} = \frac{0}{0}$$

$$\lim_{x\to 0}\frac{sen\ 2x}{sen\ 3x} = \lim_{x\to 0}\frac{\frac{d}{dx}[sen\ 2x]}{\frac{d}{dx}[sen\ 3x]}$$

$$= \lim_{x\to 0}\frac{\cos(2x)(2)}{\cos(3x)(3)}$$

When evaluating the limit the result is:

$$\lim_{x\to 0}\frac{2\cos(2x)}{3\cos(3x)} = \frac{2\cos(2*0)}{3\cos(3*0)} = \frac{2\cos(0)}{3\cos(0)} = \frac{2*1}{3*1} = \frac{2}{3}$$

As a result, you get $\qquad\boxed{\lim_{x\to 0}\frac{sen\ 2x}{sen\ 3x} = \frac{2}{3}}$

$$\lim_{x\to 0}\frac{x^3}{e^{2x}}$$

220

When evaluating the limit by direct substitution we find an indetermination of the form $\frac{\infty}{\infty}$. Therefore, the L'Hoital rule can be applied. That is, both the numerator and the denominator are derived.

$$\lim_{x\to 0}\frac{x^3}{e^{2x}}=\frac{\infty^3}{e^{2\infty}}=\frac{\infty}{\infty}$$

$$\lim_{x\to 0}\frac{x^3}{e^{2x}}=\lim_{x\to 0}\frac{\frac{d}{dx}[x^3]}{\frac{d}{dx}[e^{2x}]}=\lim_{x\to 0}\frac{3x^2}{2e^{2x}}$$

When evaluating the resulting limit, it is determined that indeterminacy persists. Then it continues to drift.

$$\lim_{x\to 0}\frac{3x^2}{2e^x}=\lim_{x\to 0}\frac{\frac{d}{dx}[3x^2]}{\frac{d}{dx}[2e^{2x}]}=\lim_{x\to 0}\frac{6x}{4e^{2x}}$$

Indeterminacy persists, it continues to derive

$$\lim_{x\to 0}\frac{\frac{d}{dx}[6x]}{\frac{d}{dx}[4e^{2x}]}=\lim_{x\to 0}\frac{6}{8e^{2x}}$$

The resulting limit is evaluated

$$\lim_{x\to 0}\frac{6}{8e^{2x}}=\frac{6}{8e^{2\infty}}=\frac{6}{\infty}=0$$

$$\boxed{\lim_{x\to 0}\frac{x^3}{e^{2x}}=0}$$

221

$$\lim_{x \to \infty} \frac{ln^2 x}{x^3}$$

When evaluating the limit by direct substitution we find an indetermination of the form $\frac{\infty}{\infty}$. Therefore, the L'Hoital rule can be applied. That is, both the numerator and the denominator are derived.

$$\lim_{x \to 0} \frac{ln^2 x}{x^3} = \lim_{x \to 0} \frac{(lnx)^2}{x^3} = \frac{(ln\infty)^2}{\infty^3} = \frac{\infty}{\infty}$$

$$\lim_{x \to 0} \frac{(lnx)^2}{x^3} = \frac{\frac{d}{dx}\left[(lnx)^2\right]}{\frac{d}{dx}[x^3]} = \lim_{x \to 0} \frac{\frac{2lnx}{x}}{3x^2}$$

$$\lim_{x \to 0} \frac{2lnx}{3x^3} = \frac{\infty}{\infty}$$

Indeterminacy persists, it continues to derive

$$\lim_{x \to 0} \frac{\frac{d}{dx}[2lnx]}{\frac{d}{dx}[3x^3]} = \lim_{x \to 0} \frac{\frac{2}{x}}{9x^2} = \lim_{x \to 0} \frac{2}{9x^3}$$

.The resulting limit is evaluated

$$\lim_{x \to 0} \frac{2}{9x^3} = \frac{2}{9*\infty^3} = \frac{2}{\infty} = 0$$

$$\boxed{\lim_{x \to \infty} \frac{ln^2 x}{x^3} = 0}$$

$$\lim_{x\to 0} \left(\frac{1}{x} - \frac{1}{sen\ x}\right)$$

When evaluating the limit by direct substitution we find an indetermination of the form $\infty - \infty$. In order to apply the L'Hopital rule, it is necessary to transform the indeterminacy $\infty - \infty$ into $\frac{0}{0}$ or $\frac{\infty}{\infty}$ This is done in the following way.

$$\lim_{x\to 0} \left(\frac{1}{x} - \frac{1}{sen\ x}\right) = \lim_{x\to 0} \left(\frac{sen\ x - x}{x\ sen\ x}\right) = \frac{0}{0}$$

Then proceed to derive numerator and denominator:

$$\lim_{x\to 0} \left(\frac{sen\ x - x}{x\ sen\ x}\right) = \lim_{x\to 0} \left[\frac{\frac{d}{dx}(sen\ x - x)}{\frac{d}{dx}(xsen\ x)}\right] \rightarrow$$

$$\rightarrow \lim_{x\to 0} \frac{\cos x - 1}{sen\ x + x\cos x} = \frac{0}{0}$$

When evaluating the limit, it is noted that the indetermination persists, then continue deriving:

$$\lim_{x\to 0} \frac{\cos x - 1}{sen\ x + x\cos x} = \lim_{x\to 0} \frac{\frac{d}{dx}(\cos x - 1)}{\frac{d}{dx}(sen\ x + x\cos x)} \rightarrow$$

$$\rightarrow \lim_{x\to 0} \frac{-sen\ x}{\cos x + \cos x - xsen\ x}$$

The resulting limit is evaluated

$$\lim_{x\to 0} \frac{-sen\ x}{\cos x + \cos x - xsen\ x} = \frac{0}{1 + 1 - 0} = \frac{0}{2} = 0$$

$$\boxed{\lim_{x\to 0} \left(\frac{1}{x} - \frac{1}{sen\ x}\right) = 0}$$

223

$$\lim_{x \to 0} \left(\frac{1}{1 + sen\ x}\right)^{\frac{1}{sen\ x}}$$

When evaluating the limit by direct substitution we find an indetermination of the form 1^{∞}. To resolve the limit, proceed as follows.

The following formula is used

$$\lim_{x \to 0} f(x)^{g(x)} = e^{\lim_{x \to x_0} g(x)(f(x)-1)}$$

Identify f (x) and g (x) and replace it in the formula

$$f(x) = \frac{1}{1+sen\ x} \qquad g(x) = \frac{1}{sen\ x}$$

$$\lim_{x \to 0} \left(\frac{1}{1+sen\ x}\right)^{\frac{1}{sen\ x}} = e^{\lim_{x \to 0}\frac{1}{sen\ x}\left(\frac{1}{1+sen\ x}-1\right)} \rightarrow$$

$$e^{\lim_{x \to 0}\frac{1}{sen\ x}\left(\frac{1-1-sen\ x}{1+sen\ x}\right)} = e^{\lim_{x \to 0}\frac{1}{sen\ x}\left(\frac{-sen\ x}{1+sen\ x}\right)} \rightarrow$$

$$= e^{\lim_{x \to 0}\frac{-sen\ x}{sen\ x+sen^2 x}}$$

When evaluating the limit, an indetermination of the form $\frac{0}{0}$

Then proceed to apply the L'Hopital rule.

$$e^{\lim_{x \to 0}\frac{-sen\ x}{sen\ x+sen^2 x}} = e^{\lim_{x \to 0}\frac{-cosx}{cosx+2senx\ cosx}} = e^{-1}$$

$$\boxed{\lim_{x \to 0} \left(\frac{1}{1 + sen\ x}\right)^{\frac{1}{sen\ x}} = e^{-1}}$$

$$\lim_{x \to \infty} \sqrt[x]{\left(\frac{x+2}{x-1}\right)^{x^2}}$$

224

The limit is rewritten and evaluated to show the type of indetermination

$$\lim_{x \to \infty} \sqrt[x]{\left(\frac{x+2}{x-1}\right)^{x^2}} = \lim_{x \to \infty} \left(\frac{x+2}{x-1}\right)^{\frac{x^2}{x}} = \lim_{x \to \infty} \left(\frac{x+2}{x-1}\right)^{x} = 1^\infty$$

Identify f (x) and g (x) and replace it in the formula (see previous problem)

$$f(x) = \frac{x+2}{x-1} \qquad\qquad y \qquad\qquad g(x) = \frac{x^2}{x}$$

$$\lim_{x \to \infty} \left(\frac{x+2}{x-1}\right)^{\frac{x^2}{x}} = e^{\lim_{x \to \infty} \frac{x^2}{x}\left(\frac{x+2}{x-1}-1\right)} \quad \to$$

$$e^{\frac{x^2}{x}\left(\frac{x+2-x+1}{x-1}\right)} = e^{\lim_{x \to \infty} \left(\frac{x^3+2x^2-x^3+x^2}{x^2-x}\right)} \quad \to \quad e^{\lim_{x \to \infty} \left(\frac{3x^2}{x^2-x}\right)}$$

When evaluating the limit, an indetermination of the form $\dfrac{\infty}{\infty}$

Then proceed to apply the L'Hopital rule

$$e^{\lim_{x \to \infty} \left(\frac{3x^2}{x^2-x}\right)} = e^{\lim_{x \to \infty} \left(\frac{6x}{2x-1}\right)} = e^{\lim_{x \to \infty} \left(\frac{6}{2}\right)} = e^3$$

$$\boxed{\lim_{x \to \infty} \sqrt[x]{\left(\frac{x+2}{x-1}\right)^{x^2}} = e^3}$$

225

$$\lim_{x \to 0^+} x^3 \cot x$$

The limit is rewritten to determine the type of indeterminacy

As the $\cot x = \dfrac{1}{\tan x}$ you have

$$\lim_{x \to 0^+} x^3 \cot x = \lim_{x \to 0^+} x^3 \frac{1}{\tan x} = \lim_{x \to 0^+} \frac{x^3}{\tan x}$$

When evaluating the limit, the indetermination turns out to be $\dfrac{0}{0}$. Therefore, the rule of L'Hopital.

$$\lim_{x \to 0^+} \frac{x^3}{\tan x} = \lim_{x \to 0^+} \frac{\frac{d}{dx}[x^3]}{\frac{d}{dx}[\tan x]} = \lim_{x \to 0^+} \frac{3x^2}{sec^2 x} = \lim_{x \to 0^+} \frac{3x^2}{\frac{1}{cos^2 x}}$$

When evaluating the resulting limit you get

$$\lim_{x \to 0^+} \frac{3x^2}{\frac{1}{cos^2 x}} = \frac{0}{1} = 0 \qquad \boxed{\mathbf{lim_{x \to 0^+} \; x^3 \cot x = 0}}$$

226

$$\lim_{x \to \infty} x \, tan \frac{1}{x}$$

When evaluating the limit the indetermination is $0 . \infty$

The limit is rewritten to transform the $0 . \infty$ en $\dfrac{0}{0}$ ó $\dfrac{\infty}{\infty}$

$$\lim_{x \to \infty} x \, tan \frac{1}{x} = \lim_{x \to \infty} \frac{tan\frac{1}{x}}{\frac{1}{x}}$$

What was done in the previous step is the application of the inverse of X.

When evaluating the limit, an indetermination of the form $\frac{0}{0}$.

Therefore, the rule of L´ Hopital.

$$\lim_{x\to\infty} \frac{\tan\frac{1}{x}}{\frac{1}{x}} = \lim_{x\to\infty} \frac{\frac{d}{dx}\left[\tan\frac{1}{x}\right]}{\frac{d}{dx}\left[\frac{1}{x}\right]} \quad \to$$

$$\lim_{x\to\infty} \frac{sec^2\left(\frac{1}{x}\right)\left(-\frac{1}{x^2}\right)}{-\frac{1}{x^2}} = \lim_{x\to\infty} sec^2\left(\frac{1}{x}\right) = \lim_{x\to\infty} \frac{1}{cos^2\left(\frac{1}{x}\right)}$$

When evaluating the limit you get

$$\lim_{x\to\infty} \frac{1}{cos^2\left(\frac{1}{x}\right)} = \frac{1}{1} = 1 \quad \boxed{\lim_{x\to\infty} x\,tan\frac{1}{x} = 1}$$

$$\boxed{\lim_{x\to 4^+} [3(x-4)]^{x-4}} \qquad \text{227}$$

When evaluating the limit, an indetermination of the form is noted 0^0.

The limit is rewritten to transform the 0^0 en $\frac{0}{0}$ ó $\frac{\infty}{\infty}$

The procedure is as follows

It is assumed that the limit exists and is equal to y.

$$y = \lim_{x\to 4^+} [3(x-4)]^{x-4}$$

Neperian is applied to both members of the equation

$$\ln y = \lim_{x\to 4^+} (x-4)\ln[3(x-4)]$$

$$\ln y = \lim_{x\to 4^+} \frac{\ln[3(x-4)]}{\frac{1}{(x-4)}} = \frac{\infty}{\infty}$$

Then proceed to derive numerator and denominator

$$\ln y = \lim_{x \to 4^+} \frac{\dfrac{3}{3(x-4)}}{-\dfrac{1}{(x-4)^2}}$$

$$= \lim_{x \to 4^+} -\frac{(x-4)^2}{(x-4)}$$

$$= \lim_{x \to 4^+} -(x-4) = 0$$

$$\ln y = 0 \to y = e^0 = 1$$

$$\lim_{x \to 4^+}[3(x-4)]^{x-4} = 1$$

$$\boxed{\lim_{x \to 4^+}[3(x-4)]^{x-4} = 1}$$

228

$$\lim_{x \to 1^+} \left(\frac{3}{\ln x} - \frac{2}{x-1}\right)$$

When evaluating the limit by direct substitution we find an indetermination of the form $\infty - \infty$. In order to apply the L'Hopital rule, it is necessary to transform the indeterminacy $\infty - \infty$ into $\dfrac{0}{0}$ or $\dfrac{\infty}{\infty}$ This is done in the following way.

The m.c.m. $(\ln x, x - 1) = (x - 1)\ln x$

$$\lim_{x \to 1^+} \left(\frac{3}{\ln x} - \frac{2}{x-1}\right) = \lim_{x \to 1^+} \left(\frac{3(x-1)-2\ln x}{(x-1)\ln x}\right)$$

$$\lim_{x \to 1^+} \left(\frac{3x-3-2\ln x}{(x-1)\ln x}\right) = \frac{0}{0}$$

Then proceed to derive numerator and denominator:

$$\lim_{x\to1^+}\left[\frac{\frac{d}{dx}(3x-3-2\ln x)}{\frac{d}{dx}((x-1)\ln x)}\right]=\lim_{x\to1^+}\left[\frac{(3-\frac{2}{x})}{\frac{(x-1)}{x}+lnx}\right]$$

When evaluating the resulting limit you get:

$$\lim_{x\to1^+}\left[\frac{(3-\frac{2}{1})}{\frac{(1-1)}{1}+ln1}\right]=\frac{1}{0}=\infty$$

$$\boxed{\lim_{x\to1^+}\left(\frac{3}{\ln x}-\frac{2}{x-1}\right)=\infty}$$

$$\boxed{\lim_{x\to0^+}\left[3(x)^{\frac{x}{2}}\right]}$$

When evaluating the limit, an indetermination of the form is noted 0^0.

The limit is rewritten to transform the $0*\infty$ en $\frac{0}{0}$ ó $\frac{\infty}{\infty}$

The procedure is as follows:

It is assumed that the limit exists and is equal to y.

$$y=\lim_{x\to0^+}\left[3(x)^{\frac{x}{2}}\right]$$

Neperian is applied to both members of the equation

$$lny=\lim_{x\to0^+}ln\left[3(x)^{\frac{x}{2}}\right]$$

$$lny=\lim_{x\to0^+}\left[\ln3+\frac{x}{2}\ln x\right]$$

$$lny=\lim_{x\to0^+}\ln3+\lim_{x\to0^+}\frac{\ln x}{\frac{2}{x}}$$

The rule is applied de L´ Hopital

$$lny = \lim_{x \to 0^+} \ln 3 + \lim_{x \to 0^+} \frac{\frac{1}{x}}{-\frac{2}{x^2}}$$

$$lny = \lim_{x \to 0^+} \ln 3 - \lim_{x \to 0^+} \frac{x}{2}$$

$$lny = \ln 3 \quad \to \quad y = 3$$

$$\boxed{\lim_{x \to 0^+} \left[3(x)^{\frac{x}{2}}\right] = 3}$$

230

$$\lim_{n \to \infty} \left(\frac{n^2 + 1}{n^2 + 3}\right)^{n}$$

When evaluating the limit, a mathematical indetermination of the form 1^{∞} ¡ do it!

Identify f (x) and g (x) and replace it in the formula

$$\lim_{x \to x_0} f(x)^{g(x)} = e^{\lim_{x \to x_0} g(x)(f(x)-1)}$$

$$f(x) = \frac{n^2+1}{n^2+3} \quad y \quad g(x) = n$$

$$\lim_{n \to \infty} \left(\frac{n^2+1}{n^2+3}\right)^{n} = e^{\lim_{n \to \infty} n(\frac{n^2+1}{n^2+3}-1)} = e^{\lim_{n \to \infty} n(\frac{n^2+1-n^2-3}{n^2+3})} =$$

$$= e^{\lim_{n \to \infty} n(\frac{-2}{n^2+3})} = e^{\lim_{n \to \infty}(\frac{-2n}{n^2+3})}$$

When evaluating the limit, an indetermination of the form $\frac{\infty}{\infty}$.

Therefore, the rule of L´ Hopital to solve the limit

$$e^{\lim_{n \to \infty}(\frac{-2n}{n^2+3})} = e^{\lim_{n \to \infty}(\frac{-2}{2n})}$$

The resulting limit is evaluated

$$e^{\lim_{n\to\infty}(\frac{-2}{2n})} = e^{\frac{-2}{2*\infty}} = e^0 = 1 \qquad \boxed{\lim_{n\to\infty}(\frac{n^2+1}{n^2+3})^n = 1}$$

$$\boxed{\lim_{x\to2^+}\left(\frac{8}{x^2-4} - \frac{x}{x-2}\right)}$$

231

When evaluating the limit by direct substitution we find an indetermination of the form $\infty - \infty$. In order to apply the L'Hopital rule, it is necessary to transform the indeterminacy $\infty - \infty$ into $\frac{0}{0}$ or $\frac{\infty}{\infty}$ This is done in the following way.

The m.c.m. $(x^2 - 4, x - 2) = x^2 - 4$

$$\lim_{x\to2^+}\left(\frac{8}{x^2-4} - \frac{x}{x-2}\right) = \lim_{x\to2^+}\left(\frac{8-x(x+2)}{x^2-4}\right)$$

$$\lim_{x\to2^+}\left(\frac{8-x^2-2x}{x^2-4}\right) = \lim_{x\to2^+}\left(-\frac{x^2+2x-8}{x^2-4}\right) = \frac{0}{0}$$

Then proceed to derive numerator and denominator:

$$-\lim_{x\to2^+}\left(\frac{2x+2}{2x}\right)$$

When evaluating the resulting limit you get:

$$-\lim_{x\to2^+}\left(\frac{2x+2}{2x}\right) = -\frac{2*2+2}{2*2} = -\frac{6}{4} = -\frac{3}{2}$$

$$\boxed{\lim_{x\to2^+}\left(\frac{8}{x^2-4} - \frac{x}{x-2}\right) = -\frac{3}{2}}$$

PACE OR RELATED REASONS FOR CHANGE

The derivative dy/dx of a function $y = f(x)$ is its instantaneous rate of change with respect to the variable **x**. When a function describes position or distance, then its rate of change with respect to time is interpreted as velocity. In general, a rate of change (or intensity of variation) with respect to time is the answer to the question **How fast does a quantity vary?** For example, if **V** represents a volume that varies or changes over time, then dv/dt is the ratio, or the rate at which the volume is changing at time **t**. A ratio of, for example, $x/dt = 10 \ cm^3/s$, means that the volume is increasing by 10 cubic centimeters every second. Similarly, if a person is walking towards the lamppost, at a constant rate of **3 feet/s**, then $dx/dt = -3$ **feet/s**. On the other hand, if the person walks away from the pole then $dx/dt = -3$ **feet/s**. The negative and positive reasons mean, of course, that the distance x is decreasing and growing, respectively.

STRATEGY TO SOLVE RHYTHM PROBLEMS OR RELATED REASONS FOR CHANGE

1] If possible, draw a diagram that illustrates the situation.

2] Designate with symbols all the quantities given the quantities to be determined that vary with time.

3] Analyze the statement of the problem and distinguish which reasons for change are known and what is the reason or pace of change that is required.

4] Present an equation that relates the variables whose reasons for change are given or have to be determined.

5] Using the chain rule, implicitly derive both members of the equation obtained in (4), with respect to time.

6] Substitute in the equation resulting from point (5), all known values of the variables and their reasons for change, in order to deduce (clear) the required rate of chan.

Pedro Pablo **CORONEL PÉREZ** / Pablo Josué **CORONEL LÓPEZ**
250 SOLVED EXERCISES FROM DERIVATIVES WITH APPLICATIONS

232

The radius *r* of a circle is growing at a rate of 3 centimeters per minute. **Calculate the rate of change of the area when a)** r_1= 6 cm and **b)** r_2= 24 cm.

Data presented by the problem

$$\frac{dr}{dt} = 3cm/min, \quad r_1 = 6cm \quad and \quad r_2 = 24cm$$

The problem asks to calculate $\dfrac{dA}{dt}$

Solution:

The formula for calculating the area of the circle is

$$A = \pi r^2$$

Using the rule of the chain and deriving both members of the equation you get

$$\frac{dA}{dt} = 2\pi r \frac{dr}{dt}$$

For $r_1 = 6cm$ you have

$$\boxed{\frac{dA}{dt} = 2\pi * 6 * 3 = 36\pi \; cm^2/min}$$

For $r_2 = 24cm$ you have

$$\boxed{\frac{dA}{dt} = 2\pi * 24 * 3 = 144\pi cm^2/min}$$

233

A spherical balloon is inflated with gas at a rate of 800 cubic centimeters per minute. **At what rate is your radio increasing at the moment it is: a)** 30 cm and **b)** 60cm?

The information that the problem gives us is

$$\frac{dV}{dt} = 800 \ cm^3/min, \qquad r_1 = 30 \ cm \quad \text{and} \quad r_2 = 60 \ cm$$

The problem asks $\frac{dr}{dt}$

The formula to calculate the volume of a sphere is: $V = \frac{4}{3}\pi r^3$

Using the rule of the chain and deriving both members of the equation you get

$$\frac{dV}{dt} = \frac{4}{3}(3)\pi r^2 \frac{dr}{dt} \quad \rightarrow \quad \frac{dV}{dt} = 4\pi r^2 \frac{dr}{dt} \quad \rightarrow \quad \frac{dr}{dt} = \frac{\frac{dV}{dt}}{4\pi r^2}$$

For $r_1 = 30 \ cm$

$$\frac{dr}{dt} = \frac{\frac{dV}{dt}}{4\pi r^2} = \frac{800}{4\pi(30)^2} = \frac{800}{3600\pi} \quad \boxed{\frac{dr}{dt} = 0.22\pi \ cm/min}$$

For $r_2 = 60 \ cm$

$$\frac{dr}{dt} = \frac{\frac{dV}{dt}}{4\pi r^2} = \frac{800}{4\pi(60)^2} = \frac{800}{14400\pi} = 0.05\pi \ cm/min$$

$$\boxed{\frac{dr}{dt} = 0.05\pi \ cm/min}$$

The formula for calculating the volume of a cone is $V = \frac{1}{3}\pi r^2 h$
Find the rate of volume change if *dr/dt* is 2 inches per minute and h=3r,

when $r_1 = 6$ inches y $r_2 = 24$ inches.

234

The information that presents us with the problem is

$$\frac{dr}{dt} = 2 \; inches/min, \quad h = 3r, \quad r_1 = 6 \text{ inches a} \quad \text{and} \quad r_2 = 24 \text{ inches}$$

The problem asks $\frac{dV}{dt}$

Using the rule of the chain and deriving both members of the equation you get

$$V = \frac{1}{3}\pi r^2 h \quad \text{For h = 3r} \quad \text{you have}$$

$$V = \frac{1}{3}\pi r^2 (3r) \quad \rightarrow \quad V = \pi r^3$$

$$\frac{dV}{dt} = 3\pi r^2 \frac{dr}{dt}$$

For $r_1 = 6$ inches

$$\frac{dV}{dt} = 3\pi r^2 \frac{dr}{dt} = 3\pi (6)^2 * 2 = 216\pi \; inches^3/min$$

$$\boxed{\frac{dV}{dt} = 216\pi \; inches^3/min}$$

For $r_2 = 24$ inches

$$\frac{dV}{dt} = 3\pi r^2 \frac{dr}{dt} = 3\pi (24)^2 * 2 = 3456\pi \; inches^3/min$$

$$\boxed{\frac{dV}{dt} = 3456\pi \; inches^3/min}$$

235

In a sand and gravel plant, the sand falls from a conveyor belt at a rate of 10 cubic feet per minute. The diameter of the base of the mound is approximately three times the height. At what rate does the heap height change when its height is 15 feet?

The information that the problem gives us is

$$\frac{dV}{dt} = 10 feet^3/min, \qquad D = 3h \qquad and \qquad h = 15 \ feet$$

The problem asks $\frac{dh}{dt}$

The formula to do the calculations is that of a cone

$$V = \frac{1}{3}\pi r^2 h \qquad For \quad 2r = 3h \quad \rightarrow \quad r = \frac{3}{2}h$$

Substituting r in the cone formula, you have

$$V = \frac{1}{3}\pi r^2 h = \frac{1}{3}\pi (\frac{3}{2}h)^2 h = \frac{1}{3}\pi \frac{9}{4}h^2 \ h$$

$$V = \frac{9}{12}\pi h^3 \rightarrow V = \frac{3}{4}\pi h^3 \ (1)$$

Note: The volume is strictly a function of h. For this reason. $r = \frac{3}{2}h.$

Using the rule of the chain and deriving both members of the equation (1)

$$\frac{dV}{dt} = \frac{9}{4}\pi h^2 \frac{dh}{dt} \qquad It \ clears \qquad \frac{dh}{dt} = \frac{4\frac{dV}{dt}}{9\pi h^2}$$

$$\frac{dh}{dt} = \frac{4\frac{dV}{dt}}{9\pi h^2} = \frac{4(10)}{9\pi(15)^2} = \frac{40}{2025\pi} = \frac{8}{405\pi}$$

$$\boxed{\frac{dh}{dt} = \frac{8}{405\pi}feet/min}$$

The combined electrical resistance R, de R_1 y R_2, connected in parallel, is given by $\frac{1}{R} = \frac{1}{R_1} + \frac{1}{R_2}$

Where R, R_1 y R_2 are measured in ohms. R_1 y R_2 are growing at a rate of 1 and 1.5 ohms per second respectively.
At what rate is R growing when R_1=50 y R_2=75 ohms respectively?

236

The information that the problem gives us is

$$\frac{dR_1}{dt} = 1 \ ohmio, \ \frac{dR_2}{dt} = 1.5 \ Ohms, \ R_1 = 50 \ ohms \ and \ R_2 = 75 \ ohms$$

The problem asks $\frac{dR}{dt}$

Using the rule of the chain and deriving both members of the combined electric equation in parallel you get

$$\frac{1}{R} = \frac{1}{R_1} + \frac{1}{R_2} \ \rightarrow \ The \ equation \ is \ rewritten: \ R^{-1} = R_1^{-1} + R_2^{-1}$$

$$-R^{-2}\frac{dR}{dt} = -R_1^{-2}\frac{dR_1}{dt} - R_2^{-2}\frac{dR_2}{dt}$$

$$-\frac{1}{R^2}\frac{dR}{dt} = -\frac{1}{R_1^2}\frac{dR_1}{dt} - \frac{1}{R_2^2}\frac{dR_2}{dt}$$

$$-\frac{1}{R^2}\frac{dR}{dt} = -\left(\frac{1}{R_1^2}\frac{dR_1}{dt} + \frac{1}{R_2^2}\frac{dR_2}{dt}\right)$$

$$\frac{1}{R^2}\frac{dR}{dt} = \left(\frac{1}{R_1^2}\frac{dR_1}{dt} + \frac{1}{R_2^2}\frac{dR_2}{dt}\right)$$

$$\frac{dR}{dt} = R^2\left(\frac{1}{R_1^2}\frac{dR_1}{dt} + \frac{1}{R_2^2}\frac{dR_2}{dt}\right)$$

From the previous equation, it is still necessary to calculate the ohmic value of R.

$$\frac{1}{R} = \frac{1}{R_1} + \frac{1}{R_2} \ \rightarrow \ R = \frac{R_1R_2}{R_1+R_2}$$

$$R = \frac{R_1R_2}{R_1+R_2} = \frac{50*75}{50+75} = \frac{3750}{125} = 30 \quad \boxed{R = 30 \ ohms}$$

Now we proceed to calculate $\dfrac{dR}{dt}$

$$\frac{dR}{dt} = R^2 \left(\frac{1}{R_1{}^2} \frac{dR_1}{dt} + \frac{1}{R_2{}^2} \frac{dR_2}{dt} \right)$$

$$\frac{dR}{dt} = (30)^2 \left(\frac{1}{(50)^2}(1) + \frac{1}{(75)^2}(1.5) \right)$$

$$\frac{dR}{dt} = 900(4 * 10^{-4} + 2.66 * 10^{-4})$$

$$\frac{dR}{dt} = 0.599 \; ohms/s$$

$$\boxed{\frac{dR}{dt} = 0.60 \; ohms/s}$$

237

A conical deposit (with the vertex down) is 10 feet wide at its highest point and is 12 feet deep (see figure). If **water is poured at a rate of 10 feet³ per minute; calculate the rate of change of water depth when it is 8 feet.**

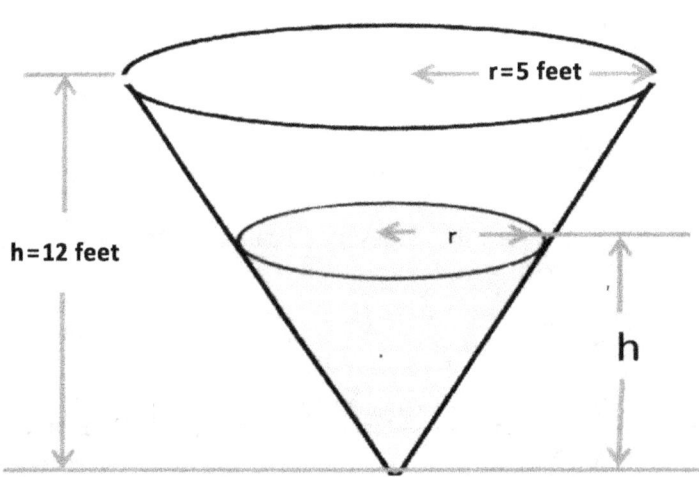

h=12 feet   r=5 feet   r   h

Information provided by the problem

$$V = \frac{1}{3}\pi r^2 h \,, \quad \frac{dV}{dt} = 10 \; feet^3/minute, \quad \frac{dh}{dt} =? \text{ and } h = 8 \; feet$$

The problem calls for the rate of change of depth $\frac{dh}{dt}$. This means that the cone volume formula must be defined only in terms of h. Therefore, it is necessary that the radius be defined in terms of h. To achieve the objective, proceed as follows: Height and radius are related (see figure).

$$\frac{h}{12} = \frac{r}{5} \quad \rightarrow \quad 5h = 12r$$

$$r = \frac{5h}{12}$$

By replacing this expression in the formula

$$V = \frac{1}{3}\pi r^2 h = \frac{1}{3}\pi \left(\frac{5h}{12}\right)^2 h$$

$$V = \frac{25\pi}{432} h^3$$

Using the rule of the chain and deriving both members of the volume equation

$$\frac{dV}{dt} = \frac{3*25\pi}{3*144} h^2 \frac{dh}{dt}$$

It clears $\frac{dh}{dt}$

$$\frac{dh}{dt} = \frac{144\frac{dV}{dt}}{25\pi h^2} \,, \quad \frac{dV}{dt} = 10 \; feet^3/minute \,, \; h = 8 \; feet$$

$$\frac{dh}{dt} = \frac{144\frac{dV}{dt}}{25\pi h^2} = \frac{144*10}{25\pi(8)^2} = \frac{1440}{1600\pi} = \frac{144}{160\pi} = \frac{9}{10\pi}$$

$$\boxed{\frac{dh}{dt} = \frac{9}{10\pi} \; feet/minute}$$

238

> An oil tank in the form of a circular cylinder with a radius equal
> to 8 meters is filling at a constant rate of **10 m³/min**.
> How fast does the oil level rise?

Data provided by the problem

$$\frac{dV}{dt} = 10m^3/min. \quad and \quad r = 8m$$

The problem asks $\dfrac{dh}{dt}$

The formula for calculating the volume of a circular cylinder is

$$V = \pi r^2 h$$

Using the rule of the chain and deriving both members of the volume equation

$$\frac{dV}{dt} = \pi r^2 \frac{dh}{dt}$$

It clears $\dfrac{dh}{dt}$

$$\frac{dh}{dt} = \frac{\frac{dV}{dt}}{\pi r^2} = \frac{10}{(8)^2\pi} = \frac{10}{64\pi} = \frac{5}{32\pi}$$

$$\boxed{\frac{dh}{dt} = \frac{5}{32\pi} \; m/min}$$

Pedro Pablo **CORONEL PÉREZ** / Pablo Josué **CORONEL LÓPEZ**
250 SOLVED EXERCISES FROM DERIVATIVES WITH APPLICATIONS

239

An insect goes along the graph of $y = x^2 + 4x + 1$, where x and y are measured in centimeters. If the **abscissa** x varies at a constant rate of 3 cm/min, how fast is the ordinate varying in (2, 13)?

Data provided by the problem

$$\frac{dx}{dt} = 3 \; cm/min$$

The problem asks $\dfrac{dy}{dx}$ on the point (2, 13)

Using the rule of the chain and deriving both members of the equation you get

$$y = x^2 + 4x + 1$$

$$\frac{dy}{dt} = 2x\frac{dx}{dt} + 4\frac{dx}{dt}$$

$$\frac{dy}{dt} = \frac{dx}{dt}(2x + 4)$$

Substituting the value of $\dfrac{dx}{dt}$ and the value of the abscissa is obtained

$$\frac{dy}{dt} = 3[2(2) + 4]$$

$$\boxed{\frac{dy}{dt} = 24 \; cm/min}$$

240

A pulley located at the top of a 12 meter building lifts a tube of the same length until it is vertical. The pulley collects the rope at a rate of -0.2 m/s. **Calculate the vertical and horizontal change rates of the end of the tube when** $y = 6$.

Free-Body diagram

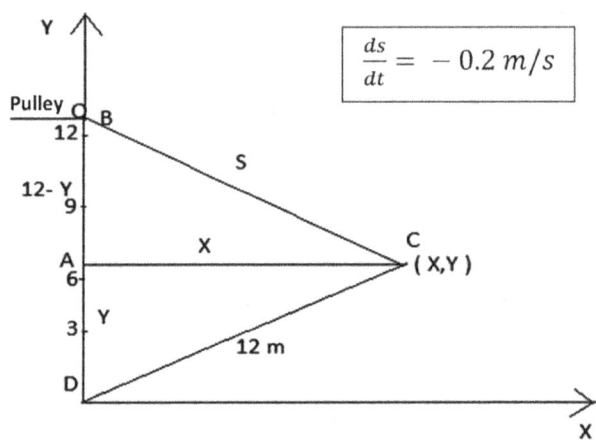

$$\frac{ds}{dt} = -0.2 \; m/s$$

Note: The AC segment is a construction that is made to divide the DBC triangle into two right triangles.

The pulley is located at point B.

At point C the end of the tube.

S identifies the rope.

Tube length: 12 m

Vertical rate of change: $\dfrac{dy}{dt}$

Horizontal rate of change: $\dfrac{dx}{dt}$

The plane geometric figures that are formed are right triangles, therefore, the Pythagorean theorem can be applied.

The Pythagorean theorem is applied to the triangle ABC

$$x^2 + (12 - y)^2 = s^2$$

Using the rule of the chain and deriving both members of the equation you get

$$2x\frac{dx}{dt} + 2(12 - y)\left(-\frac{dy}{dt}\right) = 2S\frac{ds}{dt} \quad \text{Equation 1}$$

To solve equation 1, a second equation is required, since it has two unknowns. That second equation is obtained in the following way:

The Pythagorean theorem is applied to the ACD triangle

$$x^2 + y^2 = 12^2$$

Using the rule of the chain and deriving both members of the equation you get

$$2x\frac{dx}{dt} + 2y\frac{dy}{dt} = 0$$

It clears $\frac{dy}{dt}$ Note: It could also have been cleared $\frac{dx}{dt}$

$$2y\frac{dy}{dt} = -2x\frac{dx}{dt}$$

$$\frac{dy}{dt} = \frac{-2x}{2y}\frac{dx}{dt}$$

$$\frac{dy}{dt} = -\frac{x}{y}\frac{dx}{dt} \quad \text{Equation 2}$$

Substituting equation 2 in equation 1 gives

$$2x\frac{dx}{dt} + 2(12 - y)\left(\frac{x}{y}\frac{dx}{dt}\right) = 2S\frac{ds}{dt}$$

$$2x\frac{dx}{dt} + 2(12 - y)\left(\frac{x}{y}\frac{dx}{dt}\right) = 2S\frac{ds}{dt}$$

$$2x\frac{dx}{dt} + \frac{24x}{y}\frac{dx}{dt} - \frac{2xy}{y}\frac{dx}{dt} = 2s\frac{ds}{dt}$$

$$2x\frac{dx}{dt} + \frac{24x}{y}\frac{dx}{dt} - 2x\frac{dx}{dt} = 2s\frac{ds}{dt}$$

$$\frac{dx}{dt}\left(2x - 2x + \frac{24x}{y}\right) = 2s\frac{ds}{dt}$$

$$\frac{dx}{dt}\left(\frac{24x}{y}\right) = 2s\frac{ds}{dt}$$

$$\frac{dx}{dt} = \frac{2s\frac{ds}{dt}}{\frac{24x}{y}}$$

$$\frac{dx}{dt} = \frac{2sy}{24x}\frac{ds}{dt} \qquad \boxed{\frac{dx}{dt} = \frac{sy}{12x}\frac{ds}{dt}}$$

For the calculation of $\frac{dx}{dt}$ it is previously required to calculate the values of x y s.

In the ACD triangle you get

$$x^2 + y^2 = 12^2$$

$$x = \sqrt{12^2 - y^2} = \sqrt{12^2 - (6)^2} = \sqrt{144 - 36}$$

$$x = 10.39\ m$$

In triangle ABC you get

$$s^2 = x^2 + (12 - y)^2$$

$$s = \sqrt{(10.39)^2 + (6)^2} = 11.39$$

$$s = 11.39\ m$$

Then we proceed to calculate $\frac{dx}{dt}$

$$\boxed{\frac{dx}{dt} = -0.115\ m/s}$$

$$\frac{dy}{dt} = -\frac{x}{y}\frac{dx}{dt} = -\frac{10.39}{6}(-0.115)$$

$$\boxed{\frac{dy}{dt} = 0.199\ m/s}$$

A 25-foot-long ladder is supported on a wall (SEE FIGURE). Its base slides at a rate of 2 feet per second.

a) At what rate is your top end going down the wall when the base is 7, 15, and 24 feet from the wall?

b) Determine the rate at which the area of the triangle formed by the staircase, floor and wall changes, when the base of the first is 7 feet from the wall.

c) Calculate the rate of change of the angle formed by the staircase and the wall when the base is 7 feet from the wall.

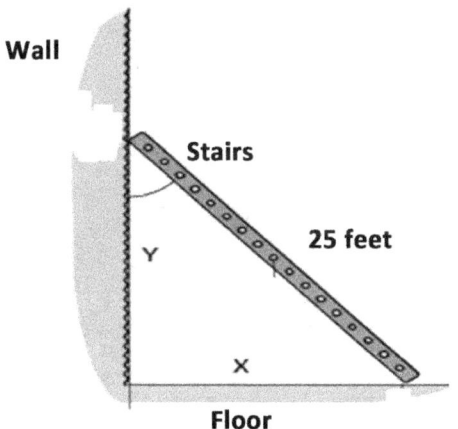

The information that the problem gives us is

Ladder length: 25 feet

Numerical values for the base: 7.15 and 24 feet.

$$\frac{dx}{dt} = 2\ feet/s$$

What is requested? $\frac{dy}{dt}, \frac{dA}{dt}, \frac{d\theta}{dt}$

Solution

a) The flat geometric figure that is formed with the staircase, the floor and the wall is a right triangle. Therefore, by applying the Pythagorean theorem you have

$$x^2 + y^2 = 25$$

Using the rule of the chain and implicitly deriving both members of the equation you get

$$2x\frac{dx}{dt} + 2y\frac{dy}{dt} = 0$$

$$2y\frac{dy}{dt} = -2x\frac{dx}{dt}$$

$$\frac{dy}{dt} = -\frac{2x}{2y} \qquad \boxed{\frac{dy}{dt} = -\frac{x}{y}\frac{dx}{dt}}$$

Calculation of y for x = 7 feet

$$x^2 + y^2 = 25$$

$$y = \sqrt{25^2 - 7^2}$$

$$y = 24\ feet$$

$$\frac{dy}{dt} = -\frac{x}{y}\frac{dx}{dt} = -\frac{7}{24} * 2$$

$$\boxed{\frac{dy}{dt} = -\frac{7}{12}\ feet/s}$$

The minus sign (-) means that the ladder slides down the wall.

Calculation of y for x = 15 feet

$$x^2 + y^2 = 25$$

$$y = \sqrt{25^2 - 15^2}$$

$$y = 20 \; feet$$

$$\frac{dy}{dt} = -\frac{x}{y}\frac{dx}{dt} = -\frac{15}{20} * 2$$

$$\boxed{\frac{dy}{dt} = -\frac{15}{10} \; feet/s}$$

Calculation of y for x = 24 feet

$$x^2 + y^2 = 25$$

$$y = \sqrt{25^2 - 24^2}$$

$$y = 7 \; feet$$

$$\frac{dy}{dt} = -\frac{x}{y}\frac{dx}{dt} = -\frac{24}{7} * 2$$

$$\frac{dy}{dt} = -\frac{48}{7} \; feet/s$$

$$\boxed{\frac{dy}{dt} = -\frac{48}{7} \; feet/s}$$

b) The formula to calculate the area of a triangle is

$$A = \frac{1}{2}bh \quad \rightarrow \quad A = \frac{1}{2}xy$$

Using the chain rule and implicitly deriving both members of the formula

$$\frac{dA}{dt} = \frac{1}{2}\left(x\frac{dy}{dt} + y\frac{dx}{dt}\right)$$

For: $x = 7$ feet $\rightarrow$ $y = 24$ feet

$$\frac{dy}{dt} = -\frac{7}{12} \; feet/s$$

$$\frac{dA}{dt} = \frac{1}{2}\left(x\frac{dy}{dt} + y\frac{dx}{dt}\right)$$

$$\frac{dA}{dt} = \frac{1}{2}\left[7\left(-\frac{7}{12}\right) + 24(2)\right]$$

$$\frac{dA}{dt} = \frac{1}{2}\left(\frac{527}{12}\right)$$

$$\boxed{\frac{dA}{dt} = \left(\frac{527}{24}\right) feet^2/s}$$

c)

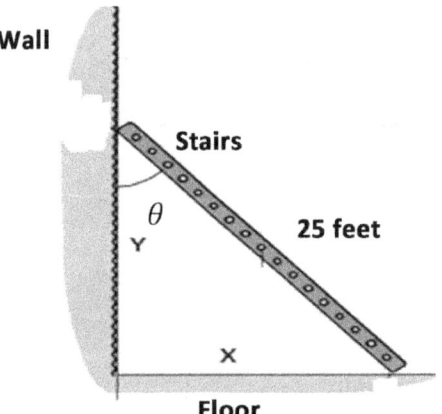

When the ladder goes down the wall it does so with a dy/dt. And when it slides it does it with a dx/dt. The trigonometric function that relates the legs is the tangent.

$$\tan\theta = \frac{x}{y}$$

Using the chain rule and implicitly deriving both members of the equation

$$sec^2\theta \frac{d\theta}{dt} = \frac{y\frac{dx}{dt} - x\frac{dy}{dt}}{y^2}$$

$$sec^2\theta \frac{d\theta}{dt} = \frac{y}{y^2}\frac{dx}{dt} - \frac{x}{y^2}\frac{dy}{dt}$$

$$sec^2\theta \frac{d\theta}{dt} = \frac{1}{y}\frac{dx}{dt} - \frac{x}{y^2}\frac{dy}{dt}$$

Como $sec^2\theta = \dfrac{1}{cos^2\theta}$ is obtained:

$$\frac{1}{cos^2\theta}\frac{d\theta}{dt} = \left(\frac{1}{y}\frac{dx}{dt} - \frac{x}{y^2}\frac{dy}{dt}\right)$$

$$\frac{d\theta}{dt} = cos^2\theta \left(\frac{1}{y}\frac{dx}{dt} - \frac{x}{y^2}\frac{dy}{dt}\right)$$

To calculate the $\dfrac{d\theta}{dt}$ it is necessary to calculate the $cos\theta$.

$$cos\theta = \frac{y}{25} = \frac{24}{25} \quad \rightarrow \quad cos\theta = \frac{24}{25}$$

For: $x = 7 feet$, $y = 24\ feet$, $\dfrac{dx}{dt} = 2\ feet/s$ and $\dfrac{dy}{dt} = -\dfrac{7}{12}\ feet/s$

$$\frac{d\theta}{dt} = cos^2\theta \left(\frac{1}{y}\frac{dx}{dt} - \frac{x}{y^2}\frac{dy}{dt}\right)$$

$$\frac{d\theta}{dt} = \left(\frac{24}{25}\right)^2 \left[\frac{1}{24} * 2 - \frac{7}{24^2}\left(-\frac{7}{12}\right)\right]$$

$$\frac{d\theta}{dt} = \frac{576}{625}\left(\frac{2}{24} + \frac{49}{6912}\right)$$

$$\frac{d\theta}{dt} = \frac{576}{625}\left(\frac{576+49}{6912}\right)$$

$$\frac{d\theta}{dt} = \frac{576}{625} * \frac{625}{6912} \quad \boxed{\frac{d\theta}{dt} = \frac{1}{12}\ rad/s}$$

OPTIMIZATION

In science, engineering and administration, it is common to be interested in the maximum and minimum values of functions; For example, a company is naturally interested in maximizing revenue while minimizing costs. The next time the reader goes to a supermarket, try this experiment: take a small ruler with you and measure the height and diameter of all the cans that contain, for example, 16 ounces of food (**28.9 *plg*^{3}**). The fact that all cans of this specified volume have the same measurements is not a coincidence, since there are specific dimensions that will minimize the amount of metal used and, therefore, minimize the manufacturing cost to the company.

In the examples and problems that follow, a function will be given, or the verbal description will have to be interpreted to establish a function from which a maximum or minimum value is sought. These are the types of verbal problems that enhance the power of calculation and provide one of the many possible answers to the old question of: **what is the use?** Next, the important steps in the solution of a problem of maximum and minimum application are indicated.

STEPS IN THE SOLUTION OF A PROBLEM WITH APPLICATION OF MAXIMUM AND MINIMUM

1] Identify all the quantities given and those to be determined. If possible, draw a drawing.

2] Write a **primary equation** for the amount that will be maximized or minimized.

3] Reduce the primary equation to one that has a single independent variable. This may involve the use of **secondary equations** that relate the independent variables of the primary equation.

4] Determine the admissible domain of the primary equation. That is, determine the values for which the problem raised makes sense.

5] Determine the maximum or minimum value desired by calculation techniques.

   Higher-order derivatives: Just as deriving a position function, a velocity function is obtained, when deriving the latter, an acceleration function is obtained. In other words, the acceleration function is the second derivative of the position function. The second derivative is an example of a higher order derivative.

A company that makes boxes wants to design an open box that has a square base and a surface area of 108 square inches, as shown in the figure. **What dimensions will the box produce with a maximum volume?**

242

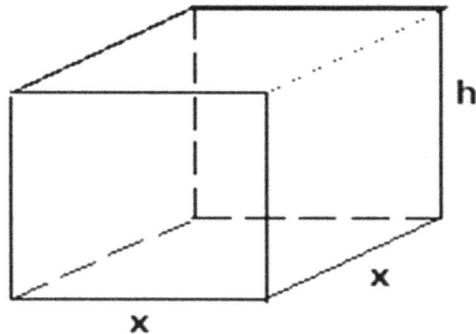

Solution

Being the base of the square box, its volume is

$$V = x * x * h$$

$$V = x^2 h \quad \textit{Equation 1}$$

This equation is called the primary equation because it provides a formula for the quantity to be optimized   Then we proceed to reduce the primary equation to one that has a single independent variable. For this it is necessary to use secondary equations that relate the independent variables of the primary equation

The surface area of the box is

S = (area of the base) + (area of the four sides)

$$S = x^2 + 4xh = 108 \quad \textit{Equation 2}$$

This equation is called the secondary equation. And it allows us to put (h) in terms of terms of x . That is to say

$$x^2 + 4xh = 108 \quad \rightarrow \quad 4xh = 108 - x^2$$

Clearing h you have

$$h = \frac{108-x^2}{4x} \quad \textit{Equation 3}$$

Now we substitute this equation into the primary equation (Equation 1) and we get

$$V = x^2\left(\frac{108-x^2}{4x}\right)$$

$$V = \frac{x(108-x^2)}{4}$$

$$V = \frac{108x-x^3}{4} = \frac{108x}{4} - \frac{x^3}{4}$$

$$\boxed{V = 27x - \frac{x^3}{4}} \quad \textit{Equation 4}$$

Before determining what values of x will produce a maximum value of V, the admissible domain needs to be determined. That is, what values of x make sense in this problem?

It is known that V≥0. Also that x must be non-negative and that the area of the base $(A = x^2)$ is at most 108.

In this way, the admissible domain is

$$0 \le x \le \sqrt{A}, \quad 0 \le x \le \sqrt{108}$$

To maximize V, the critical points of the volume function are determined (Equation 4). For this, the volume function is derived.

$$V = 27x - \frac{x^3}{4} \quad \rightarrow \quad V = 27x - \frac{1}{4}x^3$$

$$\frac{dV}{dt} = 27 - \frac{3x^2}{4} \quad \text{Equation 5}$$

Then proceed to zero the derivative

$$27 - \frac{3x^2}{4} = 0 \quad \rightarrow \quad 108 - 3x^2 = 0$$

$$3x^2 = 108 \quad \rightarrow \quad x^2 = \frac{108}{3}$$

$$x^2 = 36 \quad \rightarrow \quad x = \pm 6$$

Then the critical points are $x = \pm 6$. You do not need $x = -6$ because it is outside the domain. And now with the second derivative we verify if it is maximum or minimum. Equation 5 is derived.

$$\frac{d^2V}{dx^2} = -\frac{6}{4}x = -\frac{3}{2}x \qquad \textit{For x = 6 you get}$$

$$\frac{d^2V}{dx^2} = -\frac{3}{2} * 6 = -9 < 0 \quad \textit{Therefore, it is a maximum.}$$

Then it is concluded that V is maximum when $x = 6$. And the dimensions of the box are: $x . x . h \rightarrow 6 . 6 . 3$ inches. The height h is determined with equation 3

$$h = \frac{108 - x^2}{4x} = 3 \text{ inches}$$

243

A rectangular package that is to be sent by a postal service can have a length and a perimeter that has a maximum of 108 inches (SEE FIGURE). **Determine the dimensions of the maximum volume packet that can be sent.** (Assume that the cross section is square).

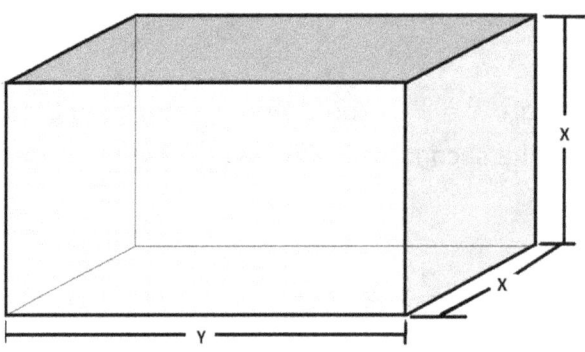

Solution

Since the cross section is square, its volume is

$$V = x^2 y \quad \textit{Equation 1}$$

This equation is called the primary equation because it provides a formula for the quantity to be optimized.

Then we proceed to reduce the primary equation to one that has a single independent variable. For this it is necessary to use secondary equations that relate the independent variables of the primary equation. The secondary equation is determined with the perimeter (P).

$$P = x + x + x + x + y \quad \textit{Equation 2}$$

$$P = 4x + y = 108$$

It clears and depending on x

$$y = 108 - 4x \quad \textit{Equation 3}$$

Substitute equation 3 in equation 1

$$V = x^2(108 - 4x)$$

$$V = 108x^2 - 4x^3 \quad \textit{Equation 4}$$

Now the admissible domain is determined

$$V \geq 0 \qquad x > 0 \qquad 0 < x < 108$$

To maximize V, the critical points of the volume function are determined (Equation 4). Deriving you have

$$\frac{dV}{dt} = 216x - 12x^2 \quad \text{Equation 5}$$

Then proceed to zero the derivative

$$216x - 12x^2 = 0 = 12x(18 - x) = 0$$

$$12x = 0 \quad \rightarrow \quad x = 0, \quad 18 - x = 0 \quad \rightarrow \quad x = 18$$

We discard the null solution because it does not make sense, and now with the second derivative we verify if it is maximum or minimum. Equation 5 is derived

$$\frac{d^2V}{dx^2} = 216 - 24x \quad \text{Para } x = 18 \text{ is obtained}$$

$$\frac{d^2V}{dx^2} = 216 - 24(18) = -216 < 0$$

It is concluded that the volume is maximum when x = 18 inches.

$$y = 108 - 4x \quad \rightarrow \quad y = 108 - 4(18) \quad \rightarrow \quad y = 36 \; inches$$

$$\boxed{x = 18 \; inches} \qquad \boxed{y = 36 \; inches}$$

The dimensions of the package are 18, 18, 36 inches.

244

A Norman window is constructed by joining a semicircle to the top of an ordinary rectangular window (SEE FIGURE). **Find the dimensions of a Norman window of maximum area if the total perimeter is 16 feet.**

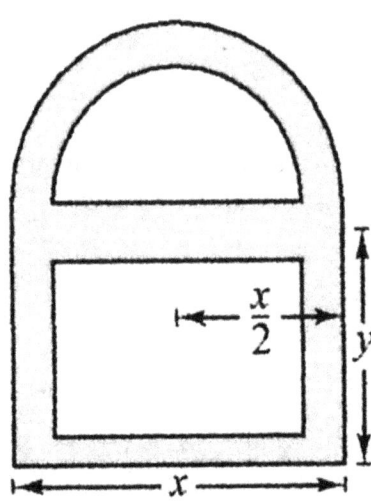

Determination of the total area (S) of the Norman window (Primary Equation).

Area of the semicircle

As the area of the circle is πr^2, that of the semicircle is then $\dfrac{\pi r^2}{2}$. Where $= \dfrac{x}{2}$.

Therefore, the area of the semicircle is $\dfrac{\pi r^2}{2} = \dfrac{\pi}{2}\left(\dfrac{x}{2}\right)^2 = \dfrac{\pi x^2}{8}$

Rectangle area: xy

Consequently, it has to be the total area (S) (Primary equation) of the Norman window is

$$S = xy + \frac{\pi x^2}{8}$$
Equation 1

Determination of the total perimeter (P) of the Norman window.

Rectangle perimeter

It consists of adding $x + y + y$. Therefore, $P = x + 2y$.

Perimeter of the semicircle

The perimeter of the circle is $2\pi r$. By dividing by two you get the perimeter of the semicircle $\dfrac{2\pi r}{2} = \pi r$. Since the radius is equal to $r = \dfrac{x}{2}$, is obtained:

Perimeter of the semicircle = $\pi \dfrac{x}{2}$

Consequently, it has to be that the total perimeter (secondary equation) is

$$P = x + 2y + \pi\frac{x}{2}$$

Then we proceed to reduce the primary equation to an equation that has a single independent variable. To do this, it is cleared and based on x in the secondary equation. Knowing that P is equal to 16 feet.

$$x + 2y + \pi\frac{x}{2} = 16$$

$$2x + 4y + \pi x = 32$$

$$4y = 32 - 2x - \pi x$$

$$y = \frac{32 - 2x - \pi x}{4} \quad \text{\textit{Equation 3}}$$

Now we substitute equation 3 in equation 1

$$S = xy + \frac{\pi x^2}{8} = x\left(\frac{32-2x-\pi x}{4}\right) + \frac{\pi x^2}{8}$$

$$S = 8x - \frac{1}{2}x^2 - \frac{\pi}{4}x^2 + \frac{\pi}{8}x^2$$

Then proceed to derive S with respect to x

$$\frac{ds}{dx} = 8 - x - \frac{\pi}{2}x + \frac{\pi}{4}x$$

$$\frac{ds}{dx} = 8 - x - \frac{\pi}{4}x$$

$$\frac{ds}{dx} = 8 - x\left(1 + \frac{\pi}{4}\right)$$

$$\boxed{\frac{ds}{dx} = 8 - x\left(1 + \frac{\pi}{4}\right)} \quad \textit{Equation 4}$$

By equating the equation 4 with zero, we obtain the critical points (maximum and minimum)

$$8 - x\left(1 + \frac{\pi}{4}\right) = 0$$

$$8 = x\left(1 + \frac{\pi}{4}\right)$$

$$x = \frac{8}{\left(1 + \frac{\pi}{4}\right)} = \frac{8}{\frac{4+\pi}{4}} \quad \boxed{x = \frac{32}{4 + \pi}}$$

The second derivative is found to determine whether it is a maximum or a minimum.

$$\frac{d^2s}{dx^2} = -\left(1 + \frac{\pi}{4}\right) < 0 \text{ Therefore, it is concluded that it is a maximum.}$$

To find the dimension and use equation 3

$$y = \frac{32 - 2x - \pi x}{4}$$

$$y = \frac{32 - 2\left[\frac{32}{4+\pi}\right] - \pi\left[\frac{32}{4+\pi}\right]}{4}$$

$$y = \frac{32 - \left[\frac{32}{4+\pi}\right][2+\pi]}{4}$$

$$y = \frac{32 - \frac{64+32\pi}{4+\pi}}{4}$$

$$y = \frac{\frac{32(4+\pi)-64-32\pi}{4+\pi}}{4}$$

$$y = \frac{\frac{128+32\pi-64-32\pi}{4+\pi}}{4}$$

$$y = \frac{\frac{64}{4+\pi}}{4} = \frac{64}{4(4+\pi)} \quad \boxed{y = \frac{16}{4+\pi}}$$

The area is maximum when

$$\boxed{x = \frac{32}{4+\pi}} \quad \boxed{y = \frac{16}{4+\pi}}$$

245

A rectangular page will contain 36 square inches of printed area. The margins on each side will be 1 1/2 inches. **Determine the dimensions of the page so that the least amount of paper is used.**

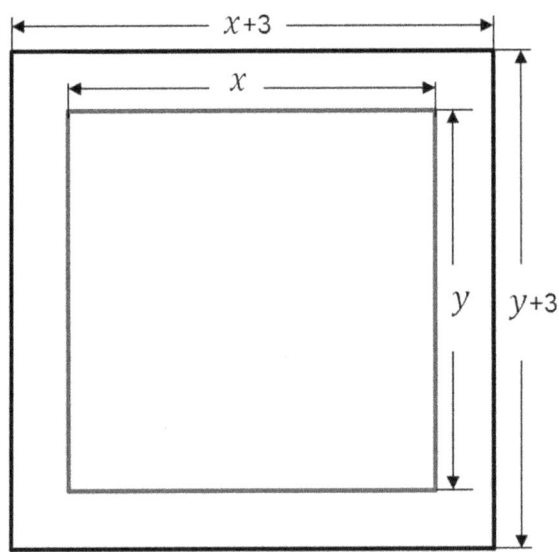

Determination of the paper area (S).

$$S = (x + 3)(y + 3) \quad \textit{Primary equation (Equation 1)}$$

Explanation: To the left and to the right of x the margin is $1\dfrac{1}{2} \rightarrow \dfrac{3}{2}$

$$x + \frac{3}{2} + \frac{3}{2} = x + \frac{6}{2} = x + 3 \quad \textit{See Figure}$$

At the top and bottom the margin is also $1\dfrac{1}{2} \rightarrow \dfrac{3}{2}$

$$y + \frac{3}{2} + \frac{3}{2} = x + \frac{6}{2} = y + 3 \quad \textit{See Figure}$$

Printed area

$$S = xy \quad \textit{Secondary equation (Equation 2)}$$

Then we proceed to reduce the primary equation to an equation that has a single independent variable. To do this, it is cleared and based on x in the secondary equation. Knowing that the printed area is 36 square inches.

$$S = xy = 36 \quad \rightarrow \quad y = \frac{36}{x} \quad \textit{Equation 3}$$

Next, equation 3 is replaced in the primary equation

$$S = (x + 3)(y + 3) = (x + 3)(\frac{36}{x} + 3)$$

$$S = \frac{36x}{x} + 3x + \frac{108}{x} + 9$$

$$S = 36 + 3x + \frac{108}{x} + 9 \quad \textit{Equation to optimize}$$

Permissible domain

$$S > 0 \qquad x > 0 \qquad 0 < x < 36$$

Next it is derived with respect to x

$$\frac{ds}{dx} = \frac{-108}{x^2} + 3 \quad \text{Equation 4}$$

By equating the equation 4 with zero, we obtain the critical points (maximum and minimum)

$$\frac{-108}{x^2} + 3 = 0 \quad \rightarrow \quad -\frac{108}{x^2} = -3$$

$$3x^2 = 108 \quad \rightarrow \quad x^2 = \frac{108}{3} = 36$$

$$x^2 = 36 \quad \rightarrow \quad x = \pm 6$$

Then the critical points are $x = \pm 6$. You do not need $x = -6$ because it is outside the domain. And now with the second derivative we verify if it is maximum or minimum. The equation 4 is derived.

$$\frac{ds}{dx} = \frac{-108}{x^2} + 3$$

$$\frac{d^2s}{dx^2} = \frac{-[-2x(108)]}{x^4} = \frac{216x}{x^4} = \frac{216}{x^3} > 0$$

Therefore, it is concluded that it is a minimum.

The solution $x = 6$ is replaced in equation 3

$$y = \frac{36}{x} = \frac{36}{6} \quad \rightarrow \quad y = 6$$

Therefore, you get

$$x = 6 \qquad y = 6$$

Consequently, to use the least amount of paper the dimensions are

$$x + 3 = 6 + 3 = 9$$

$$y + 3 = 6 + 3 = 9$$

$$\boxed{x = 9\ inches} \qquad \boxed{y = 9\ inches}$$

> Two poles, one 12 feet tall and the other 28 feet, are 30 feet away. They are supported by two cables, connected to a single stake, from ground level to the top of each pole. Where should the stake be placed so that the least amount of cable is used?

246

Data: The height of the pole AB 12 feet, the height of the pole DE 28 feet, the separation of the posts 30 feet.

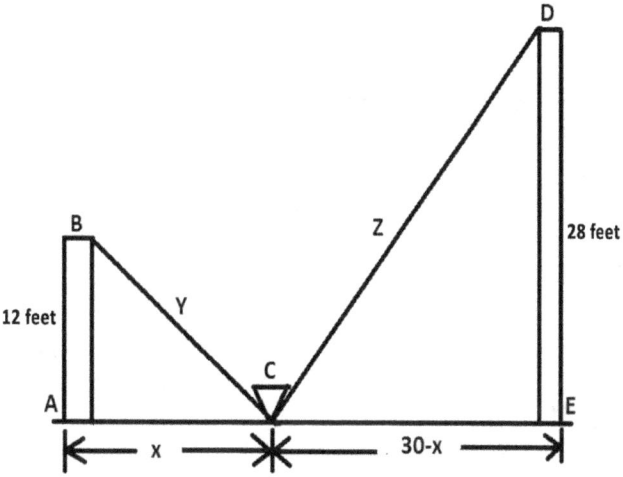

Let W be the length of the cable to be minimized. Depending on the figure you can write

$$W = y + z \ \textbf{\textit{Primary equation}}$$

In the resolution of the previous problems the variable (y) was placed according to the variable (x). In this problem, both (y) and (z) are defined as a function of a third variable (x), as can be seen in the figure.

To relate the variables x, y, z we apply the Pythagorean theorem to the ABCD and CDE triangles.

Pythagorean theorem for the ABC triangle

$$x^2 + 12^2 = y^2$$

Pythagorean theorem for the CDE triangle

$$(30 - x)^2 + 28^2 = z^2$$

Now define y and z depending on x.

$$y = \sqrt{x^2 + 144}$$

$$z = \sqrt{(30 - x)^2 + 784}$$

$$z = \sqrt{900 - 60x + x^2 + 784}$$

$$z = \sqrt{x^2 - 60x + 1684}$$

Since W is given by $W = y + z$ you have

$$w = \sqrt{x^2 + 144} + \sqrt{x^2 - 60x + 1684} \qquad \textit{Equation to minimize}$$

With a domain $0 \le x \le 30$

It is derived to W with respect to x

$$w = (x^2 + 144)^{\frac{1}{2}} + (900 - 60x + x^2 + 784)^{\frac{1}{2}}$$

$$\frac{dw}{dx} = \frac{1}{2}(x^2 + 144)^{-\frac{1}{2}}(2x) + \frac{1}{2}(x^2 - 60x + 1684)^{-\frac{1}{2}}(2x - 60)$$

$$\frac{dw}{dx} = \frac{2x}{2\sqrt{x^2+144}} + \frac{(2x-60)}{2\sqrt{x^2-60x+1684}}$$

$$\frac{dw}{dx} = \frac{x}{\sqrt{x^2+144}} + \frac{x-30}{\sqrt{x^2-60x+1684}}$$

Doing $\dfrac{dw}{dx} = 0$ and multiplying the second fraction by -1 is obtained

$$\frac{x}{\sqrt{x^2+144}} - \frac{(30-x)}{\sqrt{x^2-60x+1684}} = 0$$

The second fraction is transferred to the other member and multiplication is carried out

$$\frac{x}{\sqrt{x^2+144}} = \frac{30-x}{\sqrt{x^2-60x+1684}}$$

$$x\left(\sqrt{x^2-60x+1684}\right) = (30-x)\left(\sqrt{x^2+144}\right)$$

Squaring both members you have

$$x^2(x^2-60x+1684) = (30-x)^2(x^2+144)$$

$$x^4 - 60x^3 + 1684x^2 = (900 - 60x + x^2)(x^2 + 144)$$

$$x^4 - 60x^3 + 1684x^2 = 900x^2 + 129600 - 60x^3 - 8640x + x^4 + 144x^2$$

$$x^4 - 60x^3 + 1684x^2 = x^4 - 60x^3 + 1044x^2 - 8640x + 129600$$

$$640x^2 + 8640x - 129600 = 0$$

Solving the second degree equation gives the following values

$$x_1 = 9 \qquad\qquad x_2 = -22.5$$

The solution is discarded $x_2 = -22.5$ because it does not belong to the domain.

It should be placed 9 feet from the 12-foot pole. Or at 21 feet from the 28-foot pole.

247

The scope **R** of a projectile launched with an initial velocity V_0 at an angle θ with the horizontal is $R = \dfrac{V_0{}^2 \, sen \, 2\theta}{g}$

where **g** is the acceleration of gravity (SEE FIGURE). Determine the angle θ such that the range is a maximum.

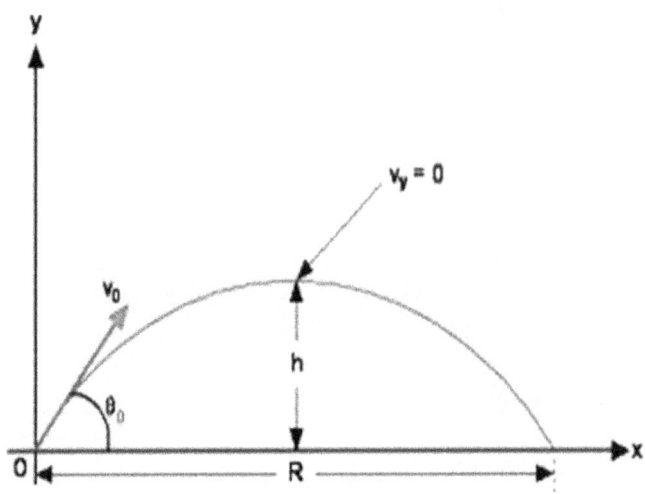

$$R = \frac{V_0{}^2 \, sen \, 2\theta}{g} \quad \textit{Primary equation}$$

Then R is derived with respect to θ

$$R = \frac{V_0{}^2}{g} \, sen \, 2\theta$$

$$\frac{dR}{d\theta} = \frac{V_0{}^2}{g} \cos 2\,\theta(2)$$

$$\frac{dR}{d\theta} = \frac{2V_0{}^2}{g} \cos 2\theta \quad \textit{Equation 1}$$

The first derivative is equated to zero (Equation 1)

$$\frac{2V_0{}^2}{g} \cos 2\theta = 0$$

$$\cos 2\theta = 0$$

For what angle θ cos2 θ = 0

$$\text{For } \frac{\pi}{4} \qquad y \qquad \frac{3\pi}{4}$$

Explanation

$$\cos 2\frac{\pi}{4} = \cos\frac{\pi}{2} = 0 \quad , \quad \cos 2\frac{3\pi}{4} = \cos\frac{3\pi}{2} = 0$$

Equation 1 is derived to determine if it is a maximum or a minimum

$$\frac{d^2R}{d\theta^2} = -\frac{4V_0{}^2}{g}\,sen\,2\theta$$

For $\theta = \dfrac{3\pi}{4}$

$$sen\,2\frac{3\pi}{4} = sen\,\frac{3\pi}{2} = -1$$

$$\frac{d^2s}{dx^2} = -\frac{4V_0{}^2}{g}\,(-1)$$

$$\frac{d^2R}{d\theta^2} > 0 \text{ It is a minimum.}$$

For $\theta = \dfrac{\pi}{4}$

$$sen\,2\frac{\pi}{4} = sen\,\frac{\pi}{2} = 1$$

$$\frac{d^2s}{dx^2} = -\frac{4V_0{}^2}{g}\,(1)$$

$\dfrac{d^2R}{d\theta^2} < 0$ It is a maximum. It is concluded that R is maximum for $\theta = \dfrac{\pi}{4}$.

248

Two factories are located at the coordinates **(-x, 0)** and **(x, 0)** with their power supply located at (0, h) (SEE FIGURE). Determine **(y)** in such a way that the total length of the electric transmission line from the power supply to the factories is minimal.

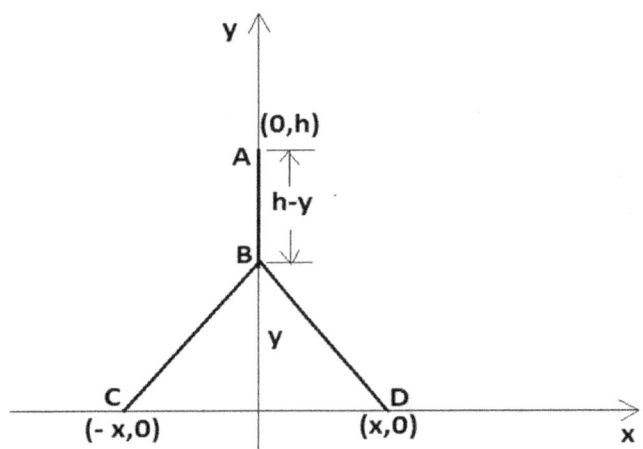

Let L be the amount of transmission line

$$L = AB + BC + BD$$

$$AB = h - y$$

$$BC = \sqrt{(-x)^2 + y^2} \quad \rightarrow \quad BC = \sqrt{x^2 + y^2}$$

$$BD = \sqrt{(x)^2 + (y)^2} \quad \rightarrow \quad BD = \sqrt{x^2 + y^2}$$

$$L = h - y + \sqrt{x^2 + y^2} + \sqrt{x^2 + y^2}$$

$$L = h - y + 2\sqrt{x^2 + y^2} \quad \textit{Primary equation}$$

Then proceed to derive L with respect to y

$$L = h - y + 2(x^2 + y^2)^{\frac{1}{2}}$$

$$\frac{dL}{dy} = -1 + 2 * \frac{1}{2}(x^2 + y^2)^{-\frac{1}{2}}(2y)$$

$$\frac{dL}{dy} = -1 + \frac{2y}{\sqrt{x^2+y^2}}$$

It is equal to zero $\frac{dL}{dy}$

$$\frac{dL}{dy} = -1 + \frac{2y}{\sqrt{x^2+y^2}} = 0$$

$$\frac{2y}{\sqrt{x^2+y^2}} = 1 \quad \rightarrow \quad 2y = \sqrt{x^2 + y^2}$$

Both members of the equation are squared

$$(2y)^2 = \left(\sqrt{x^2 + y^2}\right)^2$$

$$4y^2 = x^2 + y^2 \quad \rightarrow \quad 4y^2 - y^2 = x^2$$

$$3y^2 = x^2 \quad \rightarrow \quad y^2 = \frac{x^2}{3}$$

$$y = \sqrt{\frac{x^2}{3}} \quad \rightarrow \quad y = \frac{\sqrt{x^2}}{\sqrt{3}}$$

The amount of transmission line is minimal when $y = \dfrac{x}{\sqrt{3}}$ $\boxed{y = \dfrac{x}{\sqrt{3}}}$

249

> The potential **V** in volts at a point that is at a distance **r** from a point charge of **Q** coulombs is expressed by:
>
> $$V = \frac{KQ}{r}$$
>
> where **K** is a constant, **r** in meters.

Let Q_1 and Q_2 be two positive point charges that are separated by a distance d. (see Figure). It is requested:

(a) Deduce a mathematical expression for the potential V at a point (P) between both charges. (The potential is the sum of the one produced by each charge). Consider Q_2 = 5 Q_1

(b) Show that there is a point (P) between both charges where the potential V is minimum.

Solution

a) In the following graph the point P is represented at a distance of x meters from Q_1 and d-x meters from Q_2

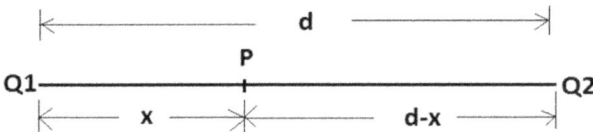

The potential V at a point P is

$$V(x) = \frac{KQ_1}{x} + \frac{KQ_2}{d-x}$$

Since Q_2 = 5 Q_1 the resulting mathematical expression is

$$V(x) = \frac{KQ_1}{x} + \frac{K5Q_1}{d-x}$$

Taking out common factor you have

$$V(x) = KQ_1 \left(\frac{1}{x} + \frac{5}{d-x}\right) \quad \textit{Primary equation}$$

b) To demonstrate that a point P exists, the domain of the function V is defined in the interval (0, d), and the primary equation is derived to obtain the critical points (maximum and minimum).

$$\frac{dV}{dx} = KQ_1 \left[\frac{-1}{x^2} + \frac{5}{(d-x)^2} \right] \rightarrow \frac{dV}{dx} = KQ_1 \left[\frac{4x^2+2dx-d^2}{x^2(d-x)^2} \right]$$

Equaling zero the first derivative you have

$$\frac{dV}{dx} = KQ_1 \left[\frac{4x^2+2dx-d^2}{x^2(d-x)^2} \right] = 0 \rightarrow 4x^2 + 2dx - d^2 = 0$$

By solving the resulting equation you get.

$$x_1 = \left(\frac{-1+\sqrt{5}}{4} \right) d \quad and \quad x_2 = \left(\frac{-1-\sqrt{5}}{4} \right) d. \text{ It discard}$$

x_2 because it is negative, $x_1 \in (o, d)$.

To determine if x_1 is a maximum or a minimum is the second derivative

$$\frac{d^2V}{dx^2} = \left[\frac{2}{x^3} + \frac{10}{(d-x)^3} \right], as \frac{d^2V}{dx^2} > 0 \rightarrow x_1 \text{ It is a minimum.}$$

The minimum voltage is then produced when the point P is at a distance x from the load Q_1 Ó at a distance (d-x) from the load Q_2.

250

A constant electromotive force generator **ε** and internal resistance **r** is connected to a load resistance **R**:

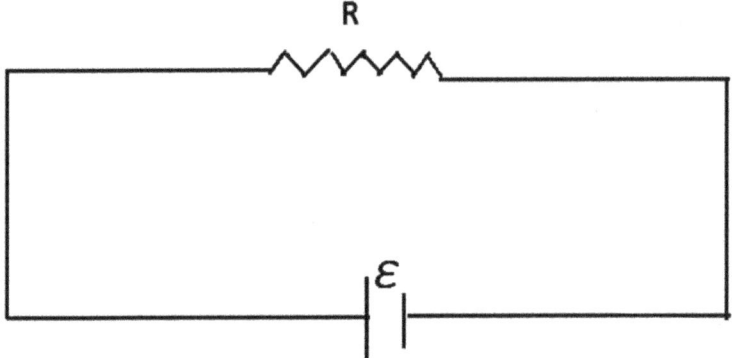

It is requested

a) Depending on the conditions of the problem, deduce a mathematical expression to determine the power P, dissipated in the load resistance.

b) Find the value of R as a function of r for the power to be maximum.

Solution

a) The intensity I of the current flowing through the given circuit is the quotient between the electromotive force ε of the generator and the total resistance of the circuit R + r.

The power dissipated in the load resistance is

$$\varepsilon, r$$

$$P = I^2 R \quad \textit{Equation 1}$$

The current in the circuit is

$$I = \frac{\varepsilon}{R} \quad \textit{Equation 2}$$

By substituting equation 2 in equation 1 and taking into account that the total resistance is $R + r$ we get

$$P = \left(\frac{\varepsilon}{(R+r)}\right)^2 R \quad \textit{Equation 3}$$

$$P = \frac{R\varepsilon^2}{(R+r)^2} \quad \textit{Primary equation}$$

To find the value of R as a function of r, we proceed to derive the primary equation, to obtain the critical points and the domain of the function P in the interval $[0, + \infty)$.

$$\frac{dP}{dR} = \varepsilon^2 \left[\frac{(R+r)^2 - 2R(R+r)}{(R+r)^4}\right]$$

$$\frac{dP}{dR} = \varepsilon^2 \left[\frac{R^2 + 2Rr + r^2 - 2R^2 - 2Rr}{(R+r)^4}\right]$$

$$\frac{dP}{dR} = \varepsilon^2 \left[\frac{-R^2 + r^2}{(R+r)^4}\right]$$

The first derivative is equal to zero

$$\frac{dP}{dR} = \varepsilon^2 \left[\frac{-R^2 + r^2}{(R+r)^4}\right] = \left[\frac{-R^2 + r^2}{(R+r)^4}\right] = 0$$

$$-R^2 + r^2 = 0 \quad \rightarrow \quad -R^2 = -r^2 \quad \rightarrow \quad R = r$$

Since the P function is continuous and positive in the interval, we can affirm that the critical point is maximum

In conclusion: "The power dissipated in the load resistance is maximum when it equals the internal resistance of the generator".

To demonstrate what is stated in the conclusion, we proceed to obtain the second derivative

$$\frac{d^2P}{dR^2} = \varepsilon^2 \left[\frac{(R+r)^4(-2R)-\left(-R^2+r^2\right)4(R+r)^3}{(R+r)^8} \right]$$

$$\frac{d^2P}{dR^2} = \varepsilon^2 \left[\frac{-2R(r+R)^4-4(-R^2+r^2)(R+r)^3}{(R+r)^8} \right]$$

$$\frac{d^2P}{dR^2} = \varepsilon^2 \left[\frac{-2(R+r)^3\left[R(R+r)+2(-R^2+r^2)\right]}{(R+r)^8} \right]$$

$$\frac{d^2P}{dR^2} = \varepsilon^2 \left[\frac{-2\left[R^2+Rr-2R^2+2r^2\right]}{(R+r)^5} \right]$$

Since R = r and operating algebraically we have

$$\frac{d^2P}{dR^2} = -\varepsilon^2 \left[\frac{4R^2}{(2R)^5} \right] \rightarrow \frac{d^2P}{dR^2} = -\varepsilon^2 \frac{4R^2}{32R^5} = -\frac{\varepsilon^2}{8R^3}$$

$$\frac{d^2P}{dR^2} = -\frac{\varepsilon^2}{8R^3}$$

Como la $\dfrac{d^2P}{dR^2} < 0$ *it is concluded then that the power in the load resistance is maximum.*

www.ingramcontent.com/pod-product-compliance
Lightning Source LLC
Chambersburg PA
CBHW081720220526
45468CB00008B/1911